Alfred Falk

Trans-Pacific sketches

a tour through the United States and Canada

Alfred Falk

Trans-Pacific sketches
a tour through the United States and Canada

ISBN/EAN: 9783744740975

Printed in Europe, USA, Canada, Australia, Japan

Cover: Foto ©Andreas Hilbeck / pixelio.de

More available books at **www.hansebooks.com**

Trans-Pacific Sketches

A TOUR THROUGH THE

United States and Canada

BY

ALFRED FALK

" With thoughts so qualified as your charities shall
Best instruct you, measure me."
Winter's Tale.

George Robertson

MELBOURNE, SYDNEY, AND ADELAIDE

MDCCCLXXVII

7056

MELBOURNE:
PRINTED BY WALKER, MAY, AND CO.,
9 MACKILLOP STREET.

To My Mother,

MRS. FANNY FALK,

THESE SKETCHES

ARE AFFECTIONATELY INSCRIBED.

January, 1877.

PREFACE.

THE following pages were written on board the good ship *Sobraon*, during her last trip from London, to Melbourne; with the sole intention and purpose of relieving the tedium, and breaking the monotony, incidental to a long sea-voyage.

The writer has been induced to publish them; it having been represented to him, that they may be of service to a large number of people, who select the American route, by which to proceed to the home country.

He has aimed at accuracy, and he thinks the facts and figures given, will generally be found correct.

He has attempted to describe American manners, and customs, impartially, and without giving offence to any citizen of the Great Republic, into whose hands this book may fall. If he has not extenuated what appeared to him as faults, he has on the other hand, given prominence to the many admirable traits, and institutions, that must be apparent to visitors to that great country ; and in conclusion he can only say, that were he not an Englishman he would wish to be an American.

A. F.

CONTENTS.

CHAPTER I.

AUCKLAND, NEW ZEALAND, AND THE FIJI ISLANDS.

CHAPTER II.

HONOLULU AND THE HAWAIIAN ARCHIPELAGO.

CHAPTER III.

SAN FRANCISCO AND SACRAMENTO CITY.

CHAPTER IV.

CROSSING THE SIERRAS—SALT LAKE CITY.

CHAPTER V.

OVER THE ROCKY MOUNTAINS—CHICAGO.

CHAPTER VI.

DETROIT—THE NIAGARA FALLS.

CHAPTER VII.

TORONTO AND TRIP DOWN THE ST. LAWRENCE.

CHAPTER VIII.

MONTREAL, QUEBEC, AND OTTAWA.

CHAPTER IX.

GENERAL REMARKS ON CANADA AND THE COLONIAL QUESTION.

CHAPTER X.

LAKES CHAMPLAIN AND GEORGE, SARATOGA, TROY, AND ALBANY.

CHAPTER XI.

THE HUDSON RIVER, WEST POINT.

CHAPTER XII.

NEW YORK CITY, BROOKLYN, AND LONG BRANCH.

CHAPTER XVI.

CINCINNATI, LOUISVILLE, AND ST. LOUIS.

CHAPTER XVII.

DOWN THE MISSISSIPPI TO NEW ORLEANS.

CHAPTER XVIII.

MOBILE, SAVANNAH, CHARLESTON, AND RICHMOND.

CHAPTER XIX.

NATIONAL CHARACTERISTICS.

CHAPTER XX.

POLITICAL CHARACTERISTICS.

TRANS-PACIFIC SKETCHES.

CHAPTER I.

AUCKLAND, NEW ZEALAND, AND THE FIJI ISLANDS.

DEPARTURE from Sydney—Three Kings' Islands—Waitemata
Harbour—Manukau Harbour—Description of Auckland—
Maories—Kandavu—Luxuriant Vegetation—Fiji Islands—
History—Natives—Dwellings—Dress—Levuka—Coral Reef—
Passing the 180th Meridian—Boat Lowering Apparatus—Boat
Accommodation of Passenger Ships—"Crossing the Line"—
Arrival at Honolulu.

ON the fourth day of April, 1876, I embarked at Sydney,
on board the good steamship *Zealandia*, for San Francisco.
Our first destination was Auckland, to receive on board
the New Zealand passengers and mails, that would otherwise have proceeded by a branch steamer, and joined us
at Kandavu, one of the Fiji Islands, had not an accident
happened to the *Colima*, the vessel intended to convey
them there.

As we steamed down Sydney harbour, its manifold
beauties were presented, like a moving panorama, to our
admiring gaze ; and again was I impressed, as I had been
on many previous occasions, with its unequalled loveliness.

It is impossible to describe that combination of beautiful
bays, and sandy coves, of lovely islets and headlands,

2

called Port Jackson; and once seen, it is equally impossible
to forget it.

As we rounded the various headlands, an entirely new
view would be disclosed, more lovely if possible, than
the preceding one; and it must have been a source of
regret to every one, that the trip down the harbour lasted
but an hour, and that we had only that time in which to
revel in its beauties and to imprint them on our minds,
to prove a source of pleasant reminiscence during our
voyage across the Pacific.

We soon steamed through the two bold rocky headlands, called the North and South Heads, the narrow
channel between which forms the entrance to Port Jackson, and were speedily on the broad Pacific, with the
Australian coast fading away in the distance.

On the morning of the fourth day out we sighted the
" Three Kings' Islands," which, seen from the distance,
seemed to be three enormous rocks, but, on a nearer
approach, proved to be a group of islands, distant about
forty miles from North Cape, the extreme northernmost
point of New Zealand. Skirting the coast, at a distance
too great to permit of our noting its general appearance,
we passed Cape Brett, situated at the entrance to the Bay
of Islands, and steaming down the Hauraki Gulf, between
the Great and Little Barrier Islands, soon entered Auckland harbour, where a remarkably pretty landscape was
spread out before us. On one side lay the city, its out-

skirts extending on either side into the hills; while the wide expanse of the harbour, with its numerous inlets and islands, lay stretched out in front of it, seemingly landlocked, and resembling a large lake, surrounded by a background of green hills. Auckland is picturesquely situated in the North Island of New Zealand, at a point where its breadth is only six miles; and possesses two fine harbours. Waitemata, the principal of these, and along the shores of which the city is built, is very fine, and one of the safest in New Zealand. It is an inlet of the Firth of Thames, and consists of an inner harbour, almost landlocked by Rangitoto and Tapu Islands, containing good wharfage accommodation ; and an outer bay of considerable extent, called Hauraki Gulf, protected by the Great and Little Barrier Islands.

On the western side of North Island, and distant only six miles from Auckland, with which it is connected, by a tramway, is Manukau harbour, on which is situated the small town of Onehunga, and from which most of the intercolonial trade is carried on.

Auckland is regularly laid out, the streets being straight and wide, and crossing one another at right angles. Queen Street, the principal thoroughfare, extends a distance of a couple of miles from the wharves, and contains the principal buildings. The public buildings are generally good, built for the most part of brick, faced with stone ; the most important being the Government

Offices, the Post Office and Custom House, and the Supreme Court. Government House is a handsome edifice, erected on an eminence which commands a grand view of the harbour, and situated in the midst of ornamental grounds, containing many fine trees.

Auckland contains something under 25,000 inhabitants. It is the second city in New Zealand, in point of population and commercial importance, and was, until 1865, the political capital of the colony; when it was found advisable to select a town with a more central position for the seat of Government, and the Legislature was removed to Wellington. Auckland has lost, in consequence, some of its former importance; but, from its unrivalled position, seems destined to become a great city, especially when the large tract of country occupied by the Maoris is brought under cultivation, and the railway system extended throughout the Island. It is at present a busy thriving place, and is rendered more characteristic of the country than other New Zealand towns by the number of the Aborigines or Maoris seen in its streets. These Maoris are a splendid race; the men being well formed, and of great size, and their features bear a great resemblance to those of Europeans, except that their noses are broader and flatter. The men dress for the most part in civilized garb, and many, notably the younger, have given up the habit of tattooing themselves. The women are pleasing in appearance, but still adhere to the practice of tattooing the lower lip. The short black pipe seen in the mouths of many of them quite destroys the romance that might otherwise attach to the softer sex of a brave and unconquered race ; for such they in reality are, living as they do, under their own king, and within their own territory, into which they will not allow any government official to enter. They are a brave race, as is proved by the fact that, during the late war, there were only 3,000 in arms against the Colonial Government, and there were at one time from 10,000 to 15,000 troops opposed to them ; and yet, in spite of this great disparity in numbers, everyone must remember how the war was protracted, and the many reverses experienced by the troops. They are shrewd and intelligent, and capable of receiving a comparatively high state of civilization, although it will be the work of several generations—that is, if they survive so long—to eradicate the savagery out of their nature. Even at the time they were first seen by the early settlers, they are described as having been more advanced than is usual with native races ; for they built themselves houses, cooked their food, and, though addicted to cannibalism, and constantly fighting amongst themselves, yet lived in communities, under their own laws, and cultivated patches of land, in which every individual of a tribe had a proprietary interest. In view of what the race is capable of, it seems a great pity that it is dying out very quickly; as from 100,000, the number they were estimated to have been in the early days of the colony, they have now decreased to 40,000, the decadence being attributed to the intemperance of the men and the unchastity of the women. I may mention that I was struck with the number of public-houses in Auckland, and on enquiry found that there are 98—in the proportion of 1 to every 250 of the inhabitants.

We were not allowed a long stay on shore; but, after embarking the New Zealand mails and passengers, we soon found ourselves again on our course, with a fair wind, and the coast of North Island disappearing on the horizon. The next four days were passed in the monotonous manner usual on shipboard; and, during the course of the fifth day, we sighted the island of Kandavu, one of the Fijian group, and in a short time were safely at anchor within its harbour. Kandavu was chosen by the mail steamers as a place of call, in preference to the other islands of the group, from its lying more directly in the track and possessing a safe harbour. It contains 124 square miles, is evidently of volcanic formation, and its appearance, as viewed from the ocean, is most lovely. Rising out of the water, to a considerable height, its slopes are covered with a dense undergrowth of green vegetation, most refreshing to eyes wearied of the eternal blue of the Pacific. Groups of graceful cocoanut palms, with their waving fronds, and beautiful green bananas, give diversity to the scene; whilst, here and there, the huts of the natives, and the modest dwellings of the white settlers may be seen peeping out from amidst the luxuriant tropical vegetation. The Island is surrounded by a coral-reef, at a distance of about three miles from the shore, and can distinctly be traced by the line of foam, caused by the sea breaking over it. The entrance to the harbour is through a passage in this reef, and the navigation is somewhat intricate and dangerous. The harbour is very safe, with good anchorage, and presents a scene of beauty that could not be anywhere exceeded. A description of Kandavu is said to be equally applicable to the other islands of the group, which bear a general resemblance to one another, varying only in size and in a few details. Having only landed on the one, I cannot, however, speak from personal observation.

The Fijian Archipelago, discovered by Tasman, the Dutch navigator, in 1643, is composed of numerous islands, of all shapes and sizes, lying midway between the Tongan Islands and the French colony of New Caledonia. The number of the islands is estimated at 225, of which 80 are inhabited. They extend nearly 300 miles from east to west, and 200 from north to south, covering an area as large as Wales. The principal are Viti Levu, Vanua Levu, Kandavu, Taviuna, and Ovalau.

The history of the islands is interesting, and may briefly be summarised as follows:—The first white settlement was made in 1804 by a party of convicts who escaped from the then existing penal establishment at Botany Bay, in New South Wales ; and the number was augmented by shipwrecked sailors and deserters from whaling ships, which often put in at the islands for water.

In 1835 a few small traders effected a lodgment at the present site of Levuka, and the small community gradually increased in numbers, until, in 1867, the white population was estimated at 500 souls. A steady tide of immigration then set in from the Australian colonies, and land was brought under cotton cultivation, for which the climate and soil were found to be well adapted. In 1873, the white population had increased to 3,000 ; the imports amounted to over £87,000, and the exports, principally of cotton, to £84,000. The unsettled state of affairs prevailing on the islands retarded their progress for some time ; but now that they have become a British colony, with a settled form of Government, they will, doubtless, enter upon an era of continuous pros-perity. The whole population, native and white, is estimated at something under 150,000.

The Fijians are, for the most part, tall and well-made, varying in colour from yellow to brown, the prevailing hue being a light brown. They are civil, orderly, and tractable; but this is solely owing to the influence and teaching of the missionaries, as they were formerly cruel, vindictive, and addicted to cannibalism. The Fijis have, in fact, been one of the most successful fields of missionary labour ; the natives now nearly all professing Christianity, and their characteristics having been beneficially modified. Their dwellings are constructed of the stalks, and thatched with the leaves of large reeds, and in appearance somewhat resemble Highland shielings. They contain no articles of furniture except mats, made of cocoanut fibre, upon which they sleep. Their dress is simple ; consisting only of a strip of cloth, or "tapa," called a sulu, wrapped round the waist, and descending to the knees. "Tapa" is a species of cloth, made from the soft bark of the paper mulberry, by beating it with wooden mallets ; and its fabrication seems to be the special work of the women. They indulge in the nasty habit of anointing their bodies with cocoa-nut oil, which makes them anything but pleasant neighbours ; they also make their hair stand up like a great mop, and wash it with lime, for the purpose of destroying the vermin. The result is, that it becomes of a bright red colour : thus a Fijian's head very much resembles a full-grown cabbage in size, and its red colour forms a remarkable and comical contrast to the black hair on his face. Levuka, the capital of the islands, is on the island of Ovalau, and is described as being built on a narrow strip of beach, from fifty to a hundred yards in width, lying at the foot of the mountains, up which the town, which at present only consists of one narrow street, will have to extend, should it, as is anticipated, increase in population and importance.

After a pleasant run on shore, we left Kandavu and proceeded on our voyage, sighting during the course of the day several other islands of the group. We obtained, too, a fine view of a coral reef, which gave us a good idea of the formation of these wonderful works of nature. The reef was of elliptical form, and extended a distance, perhaps, of ten miles, easily traceable by the line of broken water. Within, the water was placid as a mill-pond, and of a lovely green colour; forming a striking contrast to the deep blue waves of the ocean outside and the line of white foam on the reef. In this natural lake were two small islands, covered with beautiful palms and luxuriant undergrowth, completing a scene of fairy-like loveliness. These islands had all the appearance of being the tops of mountain-peaks, and had at one time, most likely, been two eminences on a larger island, which had subsided. This is the only mode of accounting for the formation of the reef, which must have been many hundred feet deep; whereas the coral insects that constructed it cannot exist at a greater depth than 120 feet. The land must, therefore, have subsided in the same ratio as the insects built upwards.

On the morning of the first day out from Kandavu we passed the 180th meridian of longitude, and, as is usual, in order to bring the ship's time more into accord with that of Greenwich, a day was added to our calendar; thus we had two Mondays, the 17th of April. On

reaching the 180th meridian our time was twelve hours
in advance of that of Greenwich, and by repeating a
day, as regards the computation of time, we practically
stood still for twenty-four hours, allowing Greenwich to
overtake and pass us ; so that we entered the Western
Hemisphere twelve hours behind it. As we travelled
west now, we had only to alter our time in accordance
with the westing made from the 180th meridian, at the
rate of four minutes to a degree, to make it agree with
that of any port into which we might put.

Selecting a fine day when we were going along with a
fair wind, the captain exercised the crew in fixing the fire-
pumps and hose, and in lowering the boats. The boats
on the davits were lowered in about a quarter of an hour ;
but it was easy to see that, had there been a heavy sea
running, as would most probably be the case in an
extremity where the boats would be requisite, they
would with great difficulty have been safely lowered into
the water. The present system of davits and falls seems
very defective ; and it appears strange that large passenger
ships are not compelled to adopt Clifford's patent falls,
which are highly commended by most nautical men.
With these falls a boat could be lowered in a couple of
minutes, and thus be the means of saving much valuable
life at sea. The boat accommodation of such ships
might, with advantage, be more stringently watched before
they leave port ; for it is notorious that the boats of most

large passenger ships are quite inadequate to contain the number of souls on board, and, in most cases, are unprovisioned and defective ; so that, small as the number of boats usually is, even these are found to be quite useless when required in any sudden emergency.

We crossed the line in longitude 161° 50', and the event was marked by a large consumption of wines and spirits, and the consequent intoxication of a number of the passengers. This was certainly no improvement on the old method of celebrating the occurrence, which, if rather rough and boisterous in its fun, was yet of an innocuous character.

On the tenth day we sighted the island of Hawaii, the largest of the Sandwich Islands, and could distinctly see the great volcanic mountain of Manua Loa, 14,000 feet high, bathed in clouds, but with its summit rising above them, clear and sharp-cut against the horizon. We soon approached the island of Oahu, upon which is the port of Honolulu ; and, taking a pilot on board, we entered the harbour in the evening, and soon found ourselves safely moored alongside the wharf. It was a feat of no small difficulty, working a long ship like the *Zealandia* alongside in the dark, and one not effected without some unpleasantness to a schooner lying at the wharf, which received a nasty hug from us as we passed, for which an action for damages has been brought against our captain.

CHAPTER II.

HONOLULU AND THE HAWAIIAN ARCHIPELAGO.

THE Sandwich Islands—Decrease of Population—Natives—Position of the Islands — Products — Government — Liquor Laws— Restriction of the Sale of Opium—Appearance of Honolulu— Harbour—Description of the Town—Taro Plant—"Poi"— Nuuanu Valley—"Pâli"—Departure from Honolulu—Arrival at San Francisco.

THE Sandwich Islands, or Hawaiian Archipelago, consist of eight larger and seven smaller islands, the principal of which are Hawaii, Maui, Kauai, and Oahu. On the latter is situated Honolulu, the capital. They contain a population of about 50,000, of which 5,000 are white settlers, chiefly British, American, and German. There is also a great number of Chinese on the islands. The present population is supposed to be little over one-tenth of what it was at the time of Captain Cook's visit in 1778; the great decrease being ascribed to the usual causes of the decadence of native races, viz., the white man's firewater and the unchastity of the women.

The natives are a fine, inoffensive race, very much resembling the New Zealand Maoris, but undisfigured by the hideous tattoo-marks—far more peacefully inclined, and more industrious; although, like most native races, they require a great incentive to labour, and consequently

do better at piece-work, than on regular wages. They all
speak more or less English, profess Christianity, and
crime is unfrequent amongst them. They vary in colour,
from yellow to a dusky brown, and many of the men
are fine big fellows, clad for the most part in European
dress, and are said to make very good sailors. The
women are pleasing in appearance, but wear an ungraceful
kind of sacque, fastened round the throat, unconfined at
the waist, and descending in long folds to their feet;
their heads are generally adorned with garlands of flowers.
Men and women are all called Kanakas, that being the
native word for man ; thus white men are Kanaka Hauri,
and natives Kanaka Mauri; but the distinctive terms in
use are Hauri and Kanaka.

The Hawaiian group lies right in the track of vessels
sailing from California to Australia and New Zealand,
and also, though less directly so, of those proceeding
from California to China and Japan. They thus form a
stepping-stone, as it were, on the road between the rising
and prosperous States on the western coast of America
and the eastern shores of Australia, and seem destined in
time, from their position, to become an important element
in the trade of the Pacific.

The products of the islands are sugar, rice, and coffee.
These constitute the exports, which in 1870 amounted
to over £400,000, whilst the imports during the same
period were nearly a like amount.

The government is vested in the king (the reigning monarch being Lunalilo), who appoints governors to the different islands to administer the laws under him; an executive council of Europeans, and a Legislative Assembly, composed in part or altogether of natives. There is a civil and criminal code, but British and American precedents are taken, and it may be said that the islands enjoy British law, modified to suit the requirements of a native race. The laws regulating the sale of liquor are very stringent, a retailer of spirituous drinks having to pay $1,000 per annum for his license, and being prohibited to supply Europeans with drink on Sunday, or natives at any time whatever, under a penalty of $500. If he cannot pay this fine, he is set to work on the coral reefs, at 25 cents a day, until the amount of the fine be worked off.

The sale of opium to the Chinese is also lessened by making it a monopoly, which is annually sold by auction. By these means the revenue has netted as much as $40,000 in a year; and the cost of the drug to the Chinese being increased by this large amount, the quantity consumed is not so great as it otherwise would be.

Honolulu, the metropolis and chief port of the group, is a pretty little town, containing about 10,000 inhabitants. Its appearance, viewed from the harbour, is remarkably picturesque, being embowered in luxuriant vegetation, with a background of volcanic hills of most irregular shapes, their sides covered with verdure, and their craggy peaks rising to a great height. In rear of the town is a peculiarly-shaped eminence called Punch-bowl Hill, the summit of which is crowned by a small battery, to protect the harbour.

The harbour consists of a roadstead and an inner basin ; the former protected by a spit of land extending into the sea, called Diamond Head, and the latter accommodating at its wharves ships of very large tonnage —our steamer, a vessel of 3,200 tons register, being very comfortably berthed. The entrance to the harbour, as in most of the islands in the Pacific, is through a passage in the coral reef that extends round the island of Oahu. These passages through the reefs are supposed to be formed by currents of cold fresh water coming from the land, in which the coral insects that build up these reefs upon submerged portions of the islands cannot exist.

Honolulu consists of regularly laid out streets, that have, however, more the appearance of wide, rural lanes, from the luxuriance of the vegetation in the gardens, and the number of beautiful trees standing between the houses. It contains some good public buildings, amongst which it is pleasing to see a public library, with 1,500 books of reference, and numerous churches. The king's palace is a plain unpretentious building of coral-stone, situated on about an acre of well laid out ground, surrounded by a high wall, at the entrance gate in which stands a native sentry, in his neat uniform of blue coat and white trousers. There are some good schools in the town, and it is a very interesting sight to see the little Kanaka children trooping off to school, with their books and slates in their hands, and satchels on their backs. There is also a quaint little theatre, situated in a pretty garden ; but there was no performance there on the only evening we were in Honolulu. I was astonished to see, in so small a place, and one where hotel accommodation is so little required, such a fine house as the Hawaiian hotel ; but I afterwards learned that it had been erected by a company formed for the purpose, and subsidized by the Government, in anticipation of the San Francisco and Yokohama mail steamers making Honolulu a place of call. It has now, I hear, been taken over entirely by the Government, and is let to its present manager free of charge, for the purpose of keeping it open.

In this equable, but enervating climate, where the thermometer registers about 80 degrees all the year round, Nature has bounteously placed the means of subsistence, without much labour, within reach of the native population, in the rapid and luxuriant growth of the taro-plant *(arum esculentum)*, from which the national dish of " poi " is made. The taro-plant is grown in shallow pits, which are kept wet, and, combined with the heat of the climate, act as forcing beds, and cause the plant to

3

grow very rapidly. The food is prepared by the succu-
lent roots being pounded into a wet pulp, which is
allowed to ferment, and is then kneaded with the hands,
until it assumes the requisite consistency; when it is
packed in calabashes, and is ready for consumption. A
taro-pit, twelve feet, square is said to produce sufficient
food, to maintain a Kanaka for a whole year. " Poi " is
eaten by inserting two fingers into the mess, twisting
them round until sufficient adheres to them, and then, by
a dexterous turn of the wrist, transferring them to the
mouth. The girls only use one finger, with which, however, they manage to make very good play.

As several of the passengers desired to see the much
talked of Pâli, a precipice situated in a gorge in the
mountains, called the Nuuanu Valley, we formed a party
for the purpose of visiting it. Taking a guide with us,
and engaging the necessary number of poor looking
horses, we started ;—passing on our way through what
may be called a suburb of the town, containing many
pretty low-roofed houses embowered in luxuriant vegetation, and standing in gardens full of splendid tropical
shade trees, such as tree ferns with stems twelve feet
high, cocoanut palms with their graceful fronds, bananas,
and tamarind trees; which gave the houses a most
refreshing appearance of coolness. Crossing a cultivated
plain two miles long, which extends from the town to the
foot of the mountains, we entered the valley and com-

menced its ascent. The road, which was very steep, continued for some distance through dense bushwood, by the side of a mountain torrent that rushed impetuously down the valley. The mountains rose on either hand to a great height, and, being covered with green foliage, presented from different points of observation a succession of grand views. After riding for some time, we were, as our guide informed us, at a distance of six miles from Honolulu, but shortly after attained the end of our up-hill journey; for, coming upon an open space, we saw the great precipice, or Pâli, in front of us, and felt ourselves well repaid by the sight for the trouble we had taken in obtaining it. It seemed to us as if, by some great convulsion of nature, half the mountain had been cut away; for, standing on the top, we had a precipice in front of us, with a sheer descent of 500 feet. The view from here was beautiful : at the base of the precipice lay a forest, beyond which lovely green country, diversified with hills and dales, extended to the coast line; whilst, out at sea, we could see the breakers on the outer reef, the white of the foam contrasting with the dark blue of the ocean, and presenting altogether a scene of rare beauty, and one not to be soon forgotten.

Returning to the town, we passed many Kanaka men and women riding furiously—that is, as furiously as their ill-bred hacks would allow them; and we were told that the natives generally have quite a passion for equestrian exercise. We noticed that the women rode astride their horses in the same manner as the men.

On our arrival in Honolulu we found, to our great regret, that the *Zealandia* was to sail in the course of a couple of hours : so, taking our last meal ashore, at the Hawaiian Hotel, and purchasing a quantity of oranges, bananas, and mangoes, we reluctantly went on board. I may mention that these oranges had a green and unripe appearance, but we found them to be most delicious, consisting of a luscious dark red pulp.

It seemed as if the departure of the mail-steamer were kept as a gala-day by the inhabitants of Honolulu ; for all the town seemed to have turned out to watch us steam out of the harbour, and we were greeted with many rounds of hearty cheers. We were soon on our course, with the coast of Oahu fading away in the distance, and with nothing left to mark the break in our journey, but a pleasant reminiscence of the agreeable short time we had spent in sunny Honolulu.

Nothing occurred to break the monotony of the voyage. About noon, on the eighth day out from Oahu, we steamed through the Golden Gate, the entrance to San Francisco Bay—past a fine battery, which commands it with its guns—thus having accomplished the 2,100 miles in seven days and a half. Steaming down the Bay, we were soon alongside the wharf; and before even the gangways were fixed, we were invaded by an army of hotel-touters, who seemed to vie with one another in giving a florid description of the particular inducements, in the way of comfort and cost of accommodation, offered by their respective hotels. It was, in fact, necessary to keep a sharp lookout over one's luggage; for some of these gentry would make a dive at a trunk, and, if not prevented, would carry it ashore. We were, however, soon landed with our *impedimenta;* and then commenced the examination of our luggage by the customs' officers, who made a careful search for contraband articles, and we had to sign declarations for even small quantities of tobacco and cigars. After having my luggage passed, I soon found myself installed in comfortable apartments in the Occidental Hotel.

CHAPTER III.

SAN FRANCISCO AND SACRAMENTO CITY.

DESCRIPTION of the City — History — Streets — Hotels — Public Buildings — Chinese Quarter — Chinese Gambling Houses — Chinese Theatres—High Cost of Commodities—Sunday Observance—Schools—Suburbs—Cliff House—Theatres—Tram Cars —Commerce—Luggage Arrangements—Description of Sacramento—Chinese Question.

SAN FRANCISCO, the chief city of the State of California, and the New York of the Pacific Coast, is situated at the end of a peninsula thirty miles long and six miles across, which separates San Francisco Bay from the Pacific Ocean. It lies at the foot of high hills, which, in the early days of the city, were cut up by numerous gullies, and the low land at their base was of small extent. Many of these hills have been levelled, the gullies filled up, and the narrow portion of the peninsula widened with land reclaimed from the ocean; thus there are now paved streets and busy thoroughfares, where once large ships rode at anchor.

The city is regularly laid out, the streets being straight, broad, and crossing one another at right angles ; the business portion is compactly built, and may be said to cover an area of nine square miles, lying between Telegraph, Rincon, and Russian Hills. It is substantially built,

for the greater part, of stone ; but the outskirts are straggling, the houses being wide apart, and generally of wood.

The rapid growth of San Francisco is almost unprecedented, equalled only by that of Melbourne, Victoria,
which owes its rise to the same cause, viz., the gold
discoveries. The first house built on the present site of
the city was in 1835, when the settlement was called
Terba Buena, from a medicinal herb found in the
vicinity ; this was changed to its present name, San
Francisco, in 1847. In 1848, when gold was first discovered, the population amounted to 1,000 ; the great
immigration from the East then set in, and in 1850 the
population had increased to 25,000 ; in 1860, to 57,000 ;
in 1870, to 150,000 ; and is now estimated at over
175,000 inhabitants.

In the early days, in consequence of the corrupt administration of the criminal laws, crimes of open violence
were of frequent occurrence, and a regular reign of terror
prevailed, until the people formed vigilance committees,
whose summary mode of procedure tended materially to
abate this state of affairs, and at the present time, life and
property are as safe, as in any other American city. I
was indeed very much struck with the quiet and orderly
state of the city after dark, and the absence of drunken
men, who, unfortunately, form such an unpleasant feature
in the streets of our own large towns ; although I was
assured by old inhabitants that drunkenness prevails to a

greater extent than in any other American city, and
that crimes of violence are much more frequent, in proportion to the population, than in New York. All I can
say is, that, if such be the case, I did not observe the
outward signs of it.

The principal thoroughfare is Montgomery-street, which,
with its fine blocks of buildings, is a very handsome
street. It extends at its northern extremity to the top of
a hill, too precipitous for carriages to ascend; but pedestrians can do so by means of a flight of steps, and from
the top a fine view of the city and bay is obtained.
Market and Kearney-streets are also fine thoroughfares,
and contain many of the best retail establishments. In
California-street are the principal banks and brokers'
offices; whilst Front, Sansome, and Battery-streets form
the centre of the wholesale trade of the city. The best
private residences are situated in Van Ness Avenue,
Pine-street Hill, Bush and Geary-streets. The traffic in
some of these streets is very great, and forcibly reminds
one of parts of the city of London.

One of the great features of San Francisco is its
enormous hotel accommodation, which would be quite
disproportionate to the requirements of the travelling
public—great as it is—were it not for the large number of
people who prefer this mode of living to having homes of
their own. Many San Franciscan ladies are averse to the
trouble of housekeeping; the reason assigned being the

difficulty they experience in procuring good domestic servants, or " helps," as they are locally called. I think, however, that in addition to this reason, it arises in a great measure from indolence, and also from the fact that, being relieved of the cares of housekeeping, they have so much more time at their disposal, to spend more pleasantly to themselves, in exhibiting their elaborate toilets in the fashionable promenades. A newly-married pair do not think of making for themselves a home of their own : they simply hire a couple of rooms at an hotel or boarding-house, and thus the comforts of home-life are quite unknown to them. Whatever reasons may be assigned for the prevalence of this system, it is radically bad, leads to much immorality, and, under it, children, being deprived of true home influence, can only grow up fast. The finest of the great hotels is undoubtedly the Palace, a colossal building, seven stories high, with a basement, occupying a whole block bounded by four streets. It is said to contain 1,800 rooms, 270 of which are bath-rooms, can accommodate about 1,300 guests, and is without doubt the largest hotel in the world. It is well managed, and is a most comfortable house to reside at ; although its great size, which makes it necessary to have five elevators to convey guests to their apartments on the upper floors, seemed to me a great drawback. Other good hotels are the Grand, the Occidental, and the Lick House.

San Francisco is very deficient in public buildings, and
of the few it possesses none have any pretensions to
architectural beauty.   There is, however, a new City Hall
in course of erection, which will, when completed, be a
very fine edifice, surpassed by few in the States.   A New
Mint is also being built, and will be a great addition
to the architecture of the city.   The Merchants' Exchange
is a fine building, and some of the banking corporations
occupy handsome premises.

A visit to the Stock Exchange is interesting, to see the
bustle and excitement, consequent on the dealing in
mining scrip.   It is customary for stock in silver mines
to be sold by auction, the bidding regulating and fixing
the prices of the day.   From the number of people
present, it seemed to me, that the whole population
must, more or less, be interested in this species of gambling—for it is nothing else.   The uproar caused by
several hundred people all speaking excitedly at the same
time completely drowns the monotonous voice of the
President, and gives to the whole proceedings the
appearance rather of a bear-garden, than of an assem-
blage of commercial men meeting together for purposes of trade.   There are several fine libraries in the
city.   The Mercantile Library occupies a fine building,
containing several spacious reading-rooms, chess-rooms,
a gallery of paintings and statuary, and 40,000 volumes.
The Oddfellows' Library numbers 25,000; the Mechanics'

Institute Library, 30,000; and the Law Library, 15,000 volumes. Thus the San Franciscans are well supplied with food for the mind ; and it must be said to their credit that these different libraries are very well patronized.

There are about 20,000 Chinese in San Francisco, and these live densely crowded together in the "Chinese quarter" in a state of filth and squalor indescribable. It seems wonderful, that they do not breed a pestilence, as the corporation does not appear to pay the least attention to the sanitary condition of this part of the city. I joined a party to visit this "quarter," and in a cellar we counted no less than fifteen Chinamen and women, living together like rabbits in a burrow, in a space that would be considered barely sufficient for two Europeans ; and a house was pointed out to us in which we were assured over 2,000 of these people were living. In the opium cellars we saw numbers of men lying in various stages of intoxication, looking haggard and blear-eyed, and showing to what depths of degradation humanity can sink. These cellars are fitted up with shelves, upon which the Chinamen lay in pairs, with boxes for pillows ; one smoking and preparing the opium, while the other lies in a state of semi-unconsciousness. We next visited one of the many gambling-houses in the "quarter," and here, in an atmosphere consisting of dense clouds of tobacco smoke, impregnated with the vilest odours imaginable, we saw a large number of Chinamen, and, to our astonishment, a few whites, sitting round tables engaged in gambling. The latter quietly slunk away at sight of the policeman, who formed one of our party. Their mode of gambling is very simple ; one throws a handful of copper coin on the table, and the others bet whether the number be odd or even. Such inveterate gamblers are the Chinese, that, at this stupid game, they will often, in one night, lose the earnings of months.

We now proceeded to one of the two Chinese theatres, but found the performance to be of a very monotonous character. The theatre itself was a low, plain building; the auditorium containing a few chairs, but the audience, for the most part, squatting on their haunches ; the stage about nine feet high, being raised a couple of feet above the ground, and ornamented all over with dirty, faded strips of red and yellow paper, printed with Chinese characters, and lighted by numerous Chinese lanterns. Pieces of tin, like sardine boxes, were piled one above the other, and with the wings, tails, and heads of birds, were nailed to the wall; whilst a miscellaneous collection of various articles, was hung all round. Amongst these articles may be enumerated old tin-pans, broken chairs, tables without legs, dirty coats and hats, rusty swords, broom handles burned black to represent spears, strips of red and yellow muslin, old boots and shoes, wooden animals painted every colour but the natural one, illustrations of junks with sails set, armies marching, and bulls fighting. The audience sat stolidly, without a smile on their faces, smoking either tobacco or opium; even the women, who occupied a compartment by themselves, indulging in this filthy habit. The orchestra sat on the stage, smoking the whole time, and amused the audience and themselves, during the performance, by clashing cymbals, beating gongs, blowing trumpets, and making generally about as unearthly a noise as it is possible to imagine. The performance consisted in several men clad in green, red, and yellow costumes, with feathers sticking out from the back of their necks, wings on their shoulders, and masks representing heads of bulls, horses, and other animals or birds, on their faces; strutting about the stage, gesticulating and shouting at one another. Part of the performance was a mimic representation of a battle, and consisted in several persons rushing on to the stage, turning somersault over the heads of some running on from the other side, and then disappearing; this was continued *ad infinitum*, and finding it to become monotonous, we did not feel inclined to wait and see if it were at all varied. These Chinese plays, we were told, generally are descriptive of the reign of a monarch, and a performance of one extends over many evenings.

In San Francisco the cost of all commodities is very high, and this in spite of the currency of the State being in gold; the farther east I proceeded, the cheaper I found the cost of everything to become, although the currency was in greenbacks, or a difference in favour of the buyer of from ten to fifteen per cent.

Owing to the large foreign population, the Sunday is observed in the Continental manner, the shops being open, and the people generally devoting the day to recreation; at the theatres the best pieces are usually performed on Sunday evening, a good audience always being reckoned upon.

The public school system of San Francisco is very good, there being a regular attendance of 30,000 children in the different school buildings. The University of California is an important educational institution, at a short distance from Oakland; and there are, in addition, a School of Design and a Medical College. The charitable institutions are numerous and excellent, and are, generally, very creditable to so young a city.

Across the bay are the pretty little towns of Oakland, Brookland, Alameda, and Saucelito, which may be regarded as suburbs of San Francisco, with which they are connected by numerous steam-ferries. They all have fine public gardens, and, being well protected from the wind by the hilly nature of the surrounding country, the climate is more equable than at San Francisco, where hot days are succeeded by cold nights. On this account, and also that the vicinity is far prettier than the sandy desert about San Francisco, many of the merchants have their residences there.

The chief point of interest in the vicinity of the city is the Cliff House, an hotel and restaurant crowning the edge of precipitous cliffs rising abruptly out of the ocean. It is distant about six miles from the city, and is reached by a fine drive; crowded, especially on Saturday afternoons, by vehicles of all descriptions. The number of high-stepping trotting horses attached to light buggies, whose drivers seem continually to aim at passing everything on the road—thus occasioning many impromptu races — gives this drive a very gay and animated appearance. In front of the hotel, are the Seal Rocks, so called from being always covered with large numbers of seals basking in the sun, and barking like dogs, and others disporting themselves in their native element. The view, from the broad piazza of this hotel, is very fine; and the Golden Gate, the entrance to San Francisco Bay, can be seen from it to great advantage.

The only public recreation ground is Golden Gate Park, which covers a large area; but, as yet, little has been done towards beautifying it. There are several squares in the city, though mostly in a neglected condition; the only exception being in the case of Portsmouth Square, which is well laid out, and surrounded by a handsome iron railing. Lone Mountain Cemetery is very beautiful, and contains some fine monuments; in it is a peculiar conical-shaped mountain, standing by itself in tolerably level country—whence its name, Lone Mountain. Its summit is surmounted by a large wooden cross, and mountain and cross form a very prominent landmark, and from its top a magnificent panorama is presented of the city, bay, and surrounding country.

San Francisco contains no less than six large theatres, the principal of which are the California Theatre, Maguire's Opera House, Wade's Opera House, and Baldwin's Academy of Music—all very fine houses, the latter especially being a perfect little bijou of a theatre, at which opera bouffe is generally performed. The arrangement of the auditorium is the same as in continental theatres, and this I found to be general throughout America ; the space that with us is used for stalls and pit being made into orchestra seats, reserved and unreserved—the latter being in front and the former at the rear, raised and separated by a barrier from the unreserved part. The price of admission is $1 to the unreserved seats, and $1.50c. to the reserved—being the same as to the dress circle.

Communication with the different parts of the city is effected, at small cost, by means of the splendid tram-car system, double tracks being laid down through every street. This is certainly the most pleasant mode of locomotion, for the roadways are paved with stones, and the jolting in ordinary vehicles is very great.

The commerce of San Francisco is very extensive, it being the great port of shipment for the whole of the

Pacific coast. Its imports consist principally of tea, coffee, sugar, rice, and coal ; its exports of gold, silver, wool, grain, and wine. The exports of the precious metals alone, in 1874, amounted to £6,000,000. It is also the centre of numerous industries, none of which, however, are as yet very extensive.

After spending a very pleasant time in San Francisco, I proceeded on my long journey East, and took the ferry-boat for Oakland, where the station, or, in American phraseology, the "depôt," of the Central Pacific Railroad is situated. We were here landed at a pier, extending two miles and a half into the Bay, where we were trans-ferred into the railway cars and started for Sacramento, where I intended to break the journey. It may be men-tioned that I found the luggage, or, as it is universally called in America, "baggage" arrangements, excellent. The trunks I did not require on the journey, I sent through to New York, receiving in lieu of them brass "checks," on presentation of which, and payment of a small charge for storage, they were delivered to me, on my arrival in that city. I took with me only what I should require on the journey, and in the same manner received checks for each package. Approaching a large town, an official walked through the cars ; and, if I intended stopping, I gave him my checks, and the name of the hotel at which I was going to stay, receiving a printed receipt in return. I then walked empty-handed to the hotel, if it were close

4

at hand, or proceeded thither by the cars, if at some
distance; thus saving the great expense of a hackney
coach.   Leaving the receipts with the clerk in the office,
I generally found my luggage deposited in my room
within half-an-hour after my arrival.

The road from San Francisco to Sacramento traverses
the valleys of the Sacramento and San Joaquin rivers,
through highly cultivated country; and passes numerous
picturesquely-situated farms, and homesteads.

Sacramento, the political capital of the " Golden State,"
is a pretty place, containing about 50,000 inhabitants.
It is situated at the head of navigation on the Sacramento
River, near its confluence with the great American river;
and is distant about 125 miles from the Pacific.   It very
much resembles a large colonial up-country town, and is
the seat of some rather extensive manufactures.   It is a
thriving and busy place, and, during the sessions of the
State Legislature, swarms with numbers of political agents
and hangers-on.   The Capitol, or State House, is a noble
building, in the Corinthian style of architecture, surmounted by a grand cupola, which renders it a very conspicuous object, from every point of the surrounding
plains.   It is, unquestionably, one of the finest edifices
in the United States.

Sacramento is the centre of a large railway system, and
from it, lines radiate to all parts of the State; hence it
seems likely in time, to become a large and important

city. The streets are broad and well paved, and lined with trees, which form an agreeable shade in summer. The houses are generally built of brick, and have a substantial and pleasing look; whilst the recent erections are really handsome in appearance.

At the time of my visit the people were much exercised in their minds about the Chinese immigration; and a petition to Congress was being got up, praying that steps might be taken to modify it. It appears that these immigrants from the Flowery Land are brought out—under the auspices of six Chinese companies, at San Francisco—under some arrangement as to labour and earnings, that tends to make the immigration very profitable to the companies; whose powers in this respect, would seem to be unlimited. The consequence is, that such numbers of Celestials are arriving in the country—there being already in San Francisco alone 20,000, equal to more than one-tenth of the whole population of that city—that the people are beginning to bestir themselves in the matter. They complain that the Chinese do not adopt the habits of the country, but live densely crowded together in their own quarters, where they become dangerous to the public health; that the gambling propensities of the men, and the prostitution of the women, are demoralizing in the extreme, more especially to the younger generation; and that, requiring as they do, so little for their maintenance, they do not compete on equal terms with Europeans, and, consequently, lower the rate of wages, and even monopolize certain trades. There is certainly a great deal of truth in these statements, quite sufficient for Congress to take the matter into consideration, and do something either to limit the immigration, or otherwise modify its pernicious effects.

Should this not be done, there will undoubtedly be a war of races ; and I " guess," Brother Jonathan, in spite of his constant mouthing about republican equality, will be very likely to " improve " his almond-eyed visitors off the face of the earth.

The Chinese themselves are afraid of this eventuality, and not long ago, in consequence of a rumour, of a rising against them, they purchased all the arms they could, in San Francisco, and stood on their defence.

CHAPTER IV.

CROSSING THE SIERRAS—SALT LAKE CITY.

SLEEPING Cars—Snow Sheds—Trestle Bridges—Shoshone Indians —Great American Desert—Ogden—Wahsatch Mountains— Salt Lake—Salt Lake City—Tabernacle—Territory of Utah— The Mormons—Their Religion—Their Account of its Origin —Gentile Account—Church Government—History—Service at the Tabernacle—Notes on Mormonism.

I LEFT Sacramento by the Central Pacific Railroad, for Ogden, the terminus of that Company's section. This journey occupies two days and nights, and would be almost insupportable, were it not for the convenience of the sleeping cars, which are most comfortable, and, being well warmed by means of stoves, tend to render endurable, at least, a trip that would otherwise be very cold and monotonous. Owing to an accident on the line, we were delayed some six hours. It is a pity that the refreshment car, at one time attached to each train, has been discontinued, as it now necessitates the passengers getting out at each station, where the train stops for refreshments.

Directly after leaving Sacramento, we commenced the ascent of the Sierra Nevada, or Snowy Range, that great tract of mountain country that forms the barrier between

the Pacific, and the Eastern States. The route lay, for
some time, through well-cultivated country; until we
arrived at Colfax, a neat and thriving little place, containing about a thousand inhabitants; where we attained
an elevation of 2,400 feet, having ascended to that height
in a distance of fifty miles. From Colfax we continued
our up-hill progress, and shortly passed along the very edge
of a tremendous chasm, 2,500 feet deep, and rounded a
bold promontory, called Cape Horn. The scenery here
is most grand and imposing, and continued so, until we
attained Summit Station, the highest point on the line, at
an altitude of 7,000 feet above the sea-level. We had
thus ascended 4,600 feet in the last fifty miles; although
we were not at the highest point of the Sierra Nevada
range, but only the elevation of the mountain-pass, which
the railroad closely follows. One cannot but appreciate,
the energy displayed in the construction of this section
of the line, which is carried along the edge of precipices
2,000 and 3,000 feet deep; and in places on narrow
ledges, which had to be excavated from the mountain side,
by men suspended from the top in baskets. Between
Colfax and Summit Stations, we passed the Great
American Cañon,—one of the grandest ravines in the
Sierra Nevada. The sides of this ravine stand like two
great perpendicular walls, each 2,000 feet high; and
between them a river dashes impetuously onward, boiling
and seething, as though in a cauldron. During our ascent

a heavy snowstorm was raging; but, thanks to our well-warmed cars, we could sit at our ease, and look out upon the wild and weird scenery.

Shortly before arriving, and after leaving Summit Station, we commenced to enter a succession of snow-sheds, erected to guard the line from avalanches of snow. These sheds—or, more correctly speaking, long wooden tunnels — are solid structures, completely covering the road for many miles, the longest of them measuring 1,700 feet, and forming a great obstruction to the view. We crossed the numerous gullies in the mountains on trestle-bridges, the frail appearance of which, at first, caused me some trepidation. They are only constructed wide enough, to allow of a single line of rail being laid, which rests upon open timbers; and crossing them is certainly quite a novel sensation, and one not unmixed with fear; for, looking out of the car windows, one does not see the track, and the train appears to be speeding through the air, with perhaps a deep ravine and mountain torrent 100 feet below.

The appearance of the Sierra generally, is that of a snow-covered country, with the various summits peeping out, like so many white mounds, covered with branching pine trees, and diversified with foaming mountain torrents.

We now commenced to descend, the track evidently following the contour of the mountains; for we seemed to wind round them in a very tortuous manner. At one time we would be whirling round the precipitous flank of a mountain, the track being suspended seemingly in the air; at another, skirting the brink of a precipice, the torrent washing its base, appearing to us like a silver thread.

After passing through most romantic scenery, we reached Truckee, a collection of wooden shanties, dignified with the appellation of " City," and dependent solely upon the lumber trade.

Truckee is at an elevation of 5,850 feet, contains about 2,000 inhabitants, and is situated on the river Truckee, in the midst of country heavily timbered with pine trees.

Continuing our course at the same high altitude, the next interesting place reached by us was Battle Mount, so called, from a sanguinary encounter, that here took place between the Indians, and the white settlers. Here we saw some of the Shoshone Indians, the original inhabitants of this part of the country, who, with their faces painted red, and their coarse black hair hanging down their shoulders, looked a very low type of humanity. The squaws, who begged money of us, doubtless for the purpose of spending in " fire-water," carried their pappooses in blankets slung across their backs. For the benefit of those who do not know what a pappoose is, I will briefly explain. A pappoose is an

Indian baby, strapped on a board about five feet in length, leather and skins of animals being nailed to it in such wise, that it resembles a huge slipper. The baby is swathed from chin to foot, its hands even being tied down ; so that nothing is visible of the living mummy but the head, which is protected by a little hood of wicker work, adorned with beads, feathers, and coloured rags. The whole apparatus is generally carried on the maternal head.

After passing Battle Mount, we traversed high plains, destitute of vegetation, except the everlasting sage-bush, and in places, the large deposits of alkali, prevented even the growth of this hardy plant. This uninteresting tract of country continued until we entered the Twelve-Mile Cañon, or Palisade. In this deep rocky ravine the bleak, broken cliffs tower on either side, whilst beneath us rolled the river, dashing up its spray as in a very frenzy, and filling the air with its sound. One of the most remarkable features of this cañon is a perpendicular mass of rock, 1,500 feet high, called the Devil's Peak. We now approached Elko, and, striking the south fork of the Humboldt River, we passed through a valley, the slopes of which were dotted with the farms of the settlers. Elko is a " city" of some 3,000 inhabitants, of rising importance, being the point where the road to the White Pine mining district branches off. Here we could see teams of mules, laden with goods, ready to start, or already on their way to Hamilton, and Treasure City, in that district. Elko is laid out in streets, and contains many stores and other buildings of a substantial character.

Leaving Elko, we soon reached the comparatively cultivated country about Toano, a small place dependent upon the mines of Eastern Nevada. This district contains the Sink of the Humboldt, a large sheet of water into which the Humboldt River, after its course of 300 miles, empties itself, but which has no outlet.

We now entered upon the Great American Desert, and the journey became very monotonous ; nothing to be seen, as far as the eye could reach, but a dreary plain, shut in by mountains, and covered with a dry kind of bush, about four inches high. This plain is sixty miles long, and as many wide ; and does not bear a living creature on its surface excepting lizards and a small animal called a jackass-rabbit. It is supposed by geologists to have been, at one time, the bed of a large inland sea. It is so thickly crusted with alkaline dust, that, from a distance, it looks as if it were covered with snow. This alkali burns the boots like lime, and the infinitesimal particles, floating in the air, irritate the throat and lungs. Altogether, we were very pleased to get out of it, and it was with a feeling of thankfulness we left it behind us.

After passing Corinne, the only little town in the territory of Utah, essentially Gentile, we arrived at

Brigham City, a Mormon settlement, which is surrounded by fruit trees, and bears a close resemblance to an English hamlet.

We now approached the Great Salt Lake—the American Dead Sea—and after skirting it for some time, came upon the cultivated country about Ogden, a clean little Mormon town of about 6,000 inhabitants, possessing a Tabernacle. At Ogden I took the cars of the Utah Central Railway Company, which at this place forms a junction with the Central Pacific line, and arrived at Salt Lake City in two hours' time; the journey being down a steep gradient, where, steam being shut off, the train proceeded at a great rate, by its own momentum only. I had expected to find a mild genial climate in Salt Lake City, arriving, as I did, in the middle of May, but was disappointed, as the whole country was still covered with snow.

The city is built at the foot of the Wahsatch Mountains, which were covered with snow from the summits to the base, and looked very imposing, seeming, as they did, to lift their white-capped tops right into the clouds. A storm in these mountains, witnessed from the city, is a grand sight. Two peaks or summits, called the Twin Peaks, distant about 15 miles south-easterly, are 11,000 feet high, and have never been free from snow since the settlement in this valley.

Salt Lake, that veritable Dead Sea, is 120 miles long, by about 45 miles broad, and contains seven islands.

Its waters are so salt that nothing can live in them, and the fish that find their way down the rivers Jordan, and Weber, are soon killed. It is said, that enough salt to supply the whole world, could be obtained by evaporation. I took a sail on the lake, but found its banks destitute of vegetation, and its dreary, barren, appearance uninviting. The rivers Jordan and Weber empty themselves into the lake, but there is no outlet; and the waters are said to have risen some 10 feet since 1850. A peculiar feature of Salt Lake is two rocks: one, a great mass, rising abruptly out of the water, and standing black and desolate, called Black Rock; the other, overhanging the margin, and bearing an indistinct resemblance to a human face, is called Profile Rock.

Salt Lake City is situated at the foot of the Wahsatch Mountains, on a plain that has, by indomitable perseverance and toil, been converted from a wilderness, impregnated with alkali, and productive only of sage-bush, into a smiling agricultural country, dotted over with the farms, or, as they are here called, the "ranches" of the settlers.

A slight description of this most remarkable city may not be out of place, before I proceed to give an account of the people who inhabit it, and of the territory in which it is situated.

Salt Lake City is regularly laid out in the form of the letter L; the larger portion stretching east and west, and the shorter north and south. The streets—as is usual in all new American towns—are wide, cross one another at right angles, and follow the cardinal points of the compass; but they possess an unique appearance, as they have brooks of clear water flowing down either side, and watering the roots of trees, planted so, as to cast a pleasant shade over the pathway. These shade-trees, bordering every thoroughfare, and the numerous orchards in, and around the city, give it the appearance of being embowered in foliage—very refreshing to the eye in the hot season. The city is laid out in square blocks of ten acres each, covers a space of about nine square miles, and contains a population of some 25,000 inhabitants; 20,000 of whom are Mormons, and the remainder Gentiles—as all are called who do not profess the faith of Mormon.

Prominent amongst the principal buildings is the Tabernacle, an oval building, 250 feet long, by 150 feet wide, and 80 feet high, built on stone pillars 20 feet high, the roof being a lattice-work of red pine, unsupported by a single column. From its peculiar form it is a prominent feature in the city, and can be recognized from every part of it, by its egg-shaped, dome-like roof. It has a gallery running round three sides of the building, and will seat from 10,000 to 12,000 people. The acoustic properties are splendid; and it would form a good model on which to build music and lecture halls. A person

speaking in his usual tone of voice, can be heard all over
the large building.   It possesses the third largest organ
in the United States, the tone of which is very fine ; and
it is noteworthy, that this organ, with its front towers 58
feet high and its gilded pipes 38 feet long, was entirely
built in the city.

The other principal edifices are the Courthouse, the
University, the City Hall, and the Theatre.   The new
Temple, now in course of erection, is intended to be on
a colossal scale ; it has been many years in building, and
will not be completed for many years to come.   It forms
the centre of the hopes of the many thousand devotees
of Mormonism, who seem to regard its erection as an
article of faith.   It is to be devoted to such preliminary
rites and ceremonies as baptisms, washings, anointings,
etc.

The Territory of Utah, of which Salt Lake City is the
capital, contains a population of 86,605 ; of whom
about 60,000 may be Mormons, the remainder being, for
the most part, miners attracted by the great mineral
wealth of the Territory—rich veins of gold, silver, iron,
and other metals having been discovered.   .

By the last census, it is seen that the males exceed
the females by 1,277 ; but it must be borne in mind that,
in ordinary cases, in newly settled countries, the males
much more largely outnumber the females.   The returns
for Salt Lake City also show how largely the " peculiar

institution" of the Mormons is sustained by the foreign portion of the community. The native-born population number 10,236, and the males exceed the females by 78; the foreign-born population are 7,010 in number, and the females exceed the males by 686. Thus, in the native population of Salt Lake City, the proportion is 50 females to 51 males; and in the foreign, 38 females to 31 males. If children, who are probably in about equal proportions, be excepted, the excess of women over men becomes more marked.

Utah, thirty years ago a desert, is now a land of industry and wealth; its soil teeming with riches, and supporting a large population, who enjoy in peace, the products of their labour. Prosperous little towns, and villages, extend over a distance of 500 miles; and 220 schools provide for the mental cultivation, of the rising generation.

Utah, like other territories in the Union, returns one member to Congress, who has the privilege of taking part in a debate, but not the power of voting. It contains thirty incorporated towns; and the government is vested in a Governor, chosen by the President of the United States, a Council of thirteen members, and a House of Representatives of twenty-six members, both being elected for two years, and the sessions being biennial. Copies of all laws passed by the Assembly, and signed by the Governor, are forwarded to the presiding officers of both Houses of Congress; and if disapproved by that body, become null and void. The Governor is elected for four years; and Brigham Young, President of the Church of Jesus Christ of Latter-Day Saints, once occupied the position. The United States maintain a garrison of 1,700 men, at a distance of three miles from the city, at a remarkably picturesque place, called Camp Douglas.

The judicial power of the Territory consists of a Supreme Court, District, and Probate Courts, and Justices of the Peace. The Supreme Court is presided over by the Chief Justice and two associate Judges, all appointed by the President of the United States. The Territory is divided into three districts, one of the Judges of the Supreme Court being allotted to each as District Judge. Justices of the peace have no jurisdiction in cases involving sums of over $100, or in questions of boundaries, or titles of land. A Probate Judge is elected for each county, by the Legislative Assembly. He holds office for four years, and has civil, criminal, and surrogate jurisdiction in cases arising in the county. Appeals may be taken from the Probate to the District Court, and thence to the Supreme Court. Each county elects in addition, for three years, three select men, who, with the Probate Judge, form a County Court, whose business is to divide the county into precincts or municipalities, school districts, roads, boundaries of irrigation districts, to levy taxes for the erection and maintenance of county buildings, and provide for pounds for stray cattle, &c. It has been attempted to get the Territory of Utah incorporated as a State in the Union, under the name of the " State of Deseret," but polygamy stands in the way.

Socially, the people are hard-working and industrious, as is proved by the fact that the first pioneers arrived in these vallies in 1847, and they have, in thirty years, transformed what was then a barren desert into a productive country, giving homes to thousands. They were certainly fortunate in their choice of country; for, although the soil, impregnated as it was with alkali, did not yield much for the first few years, yet, after being turned over several times, this very alkali in the soil served as an element of richness. The early settlers, too, when the gold discoveries in California took place, were right in the track of the miners and others travelling overland to the El Dorado, and reaped great profit out of them.

They profess the faith of Mormon, as revealed to their Prophet, Joseph Smith, under the style of the " Church of Jesus Christ of Latter-Day Saints." The following is the Mormon account of the origin of their belief :—Joseph Smith, the founder, was born at Sharon, Vermont, in 1805; he was an early seeker after knowledge, and was rewarded by having a vision. In this vision he saw two Celestial Beings—the Father and the Son—who took the trouble to inform him, that all existing Faiths were

5

incorrect ; that the Covenant once made with Israel was near fulfilment ; and that he had been chosen the instrument to prepare the people for the second advent of the Saviour—the millennium being close at hand. At a later period he had another vision ; this time seeing the Angel Moroni, who delivered to him a series of metal plates, of the appearance of gold, eight inches long and six inches wide, bound together like the leaves of a book by three rings ; thus forming a volume six inches thick, written in Egyptian characters. He received at the same time the Urim and Thumim, two transparent stones, set in the two rims of a bow, and used in the same manner as spectacles. By the help of these he translated the records, and learned that the American continent was first colonized by a people who came from the Tower of Babel, after the confusion of tongues, who were called Jaredites ; and also by a colony direct from Jerusalem, about six centuries B.C. ; these latter being Israelites, the descendants of Joseph. He also learned that the Saviour appeared in America after His Resurrection ; that He planted there the Gospel in all its fulness; that there were Apostles, Prophets, Pastors, Teachers, and Evangelists, and the same ordinances were enjoyed as in the Eastern Hemisphere ; that the people were cut off because of their great transgressions ; and that their last prophets were commanded to write on metal plates an abridgment of their prophecies, history, &c., and to hide them in the earth, until such time as they should be brought forth and be united to the Bible, for the accomplishment of God's purposes in the latter days. The records proceeded to tell how the Jaredites were destroyed about the time the Israelites arrived in the country ; and how the latter became divided into two nations, the Nephites, or followers of Nephi, the inspired writer of the metal plates ; and the Lamanites, or unbelievers, who, for their sins, were condemned to have red skins, and to become " an idle people, full of mischief and subtlety," and whose descendants are the present American Indians. How the Nephites fell in battle against the Lamanites, towards the close of the fourth century, and how they buried the plates, as instructed, at Mount Cumorah, where they made their final unsuccessful stand. The translation of these records, made by Joseph Smith, forms the Book of Mormon, or Golden Bible, as it was at first called, upon which is founded modern Mormonism.

This Mormon account of the foundation of the belief might have remained unquestioned had that farrago of ancient Jewish ceremonies, grafted upon a spurious kind of Christianity, called Mormonism, remained within its original limits, especially as its professors were a harmless, hard-working class, and the Book of Mormon inculcated morality and specially prohibited polygamy. But as the numbers of converts increased, so the Revelation of

Celestial Marriage was introduced ; and its debasing effects converted the members of the new sect into law-breakers, and made them a cancer and sore in the heart of the civilized communities amongst which they dwelt, and thus awakened that feeling of animosity and open hostility against which they have had to contend.

In juxtaposition, therefore, to the Mormon account given above, I will place the Gentile version, as proved by researches made. The Smith family is said, in the sworn testimony of sixty of the most respectable citizens of Wayne County, to have been " false, immoral, and fraudulent," and Joseph, " the worst of the whole." Whatever value may attach to this sworn testimony, Joseph Smith, on the showing of Orson Pratt, the Mormon Bishop, led a most dissolute and disreputable life. To distinguish, however, whether he was a religious enthu-siast or an impostor, it is necessary to examine the Gentile version of the episode connected with the finding and translation of the Book of Mormon. It is to the effect that Solomon Spaulding, born at Ashford, Con-necticut, in 1761, a graduate of Dartmouth College, who had formerly been a minister of religion, but who after-wards went into trade and became insolvent, wrote a religious novel, entitled " The Manuscript Found," based upon the theory then prevalent that the Indians were the descendants of the ten lost tribes of Israel ; and that to this production Smith was indebted for his Book of

Mormon. The evidences in favour of such being the case are very strong, the plot and characters being the same in both ; the difference between them being only the addition in Smith's book of some ungrammatical religious matter. The leading characters in both are Mormon and his son Moroni, Lehi, Nephi, and the Lamanites. It was only in the latter end of the year 1827 that Smith professed to have received the plates, and as early as 1813 "The Manuscript Found" was advertised in the papers as shortly forthcoming, " with a full account of the Book of Mormon." In 1812, it had been placed in the hands of a Mr. Patterson, a printer, in Pittsburg; but before any arrangements could be made for its publication Spaulding died, the MSS. remaining in Patterson's possession. It is supposed that a copy was made, or the original MSS. stolen, by one Sidney Rigdon, a compositor in Patterson's office, who afterwards joined the Mormons and became a prominent man amongst them.

Smith's mode of procedure, when translating the " plates," was certainly suspicious. A blanket was suspended across the room, to conceal the sacred records from profane and prying eyes, and, sitting behind this screen, by the help of the spectacles before mentioned, he dictated to Oliver Cowderoy, his coadjutor, the suppositious translation. The work, when completed, before publication, required corroboration, and three

witnesses were accordingly "raised up" in the persons
of Oliver Cowderoy, David Whitmer, and Martin Harris
—all of whom afterwards quarrelled with Smith, and, in
consequence, apostatized, when they denied the genuine-
ness of the records.   Eight other witnesses were afterwards found, who made declarations that they had seen
and handled the metal plates translated by Smith.   All
these eight persons, however, belonged to two families
only, and three of them were the father and two brothers
of Smith.   With these exceptions, no person, either
Mormon or Gentile, has seen these records, and all
knowledge of them is quite traditional.   No sooner was
the Book of Mormon published than the widow and
brother of Spaulding identified it with the novel of "The
Manuscript Found;" and several others, who had heard
portions read, did the same.   All these facts tend to
prove that the whole was a fraud on the part of Smith,
for the purposes, it is to be presumed, of gain and
notoriety.

Finding it necessary to confirm his power on a sure
basis, and to form a proper church government for the
increased numbers of the new sect, Smith declared that
the Saviour, Moses, Elias, and Elijah had appeared to
himself and Cowderoy, and had delivered into their
hands the keys of the various priesthoods, and unlimited
spiritual and temporal power; and that from St. John
the Baptist they had received the powers of baptism.

In consequence, the pure priesthoods, Melchizedek and Levitical—which had existed in their purity in the time of Moses, and which, after degeneration, had been revived by Christ in the persons of the Apostles—were again to be resuscitated in their most perfect form by Smith, with the assistance of these later revelations.

The church government, gradually built up after the receipt of these powers, was divided into the two branches—Melchizedek and Levitical—as enjoined, the latter presided over by the Bishop, and is at the present time vested in the following officers :—the President of the Church, who is assisted in his deliberations by two councillors, also called Presidents ; the Patriarch, who is the second officer in point of dignity ; the twelve Apostles, called the Council of Twelve, whose duties are to ordain all elders, priests, teachers, and deacons, and to baptize and administer the sacrament ; the Quorum of Seventy, who are under the direction of the " twelve," and who form the missionaries and preachers of the sect ; the High Priests, who are men advanced in years, and who officiate in the offices of the church, in the absence of the higher authorities ; the elders, who preside at meetings, and exercise a general supervision over the priests ; the teachers, who are assistants to the priests ; and the deacons, who act as church collectors, and perform minor offices. The Council of Twelve, the Quorum of Seventy, the Patriarch, High Priests, and elders, belong to the Melchizedek order; the Bishop, priests, teachers, and deacons, to the Levitical. The duties of the latter order are to attend to the work of the Temple, and its members are chosen from the " lineal descendants of Aaron," who are pointed out by special revelation.

Every lay member of the community pays, either in money or kind, a tithing, for the receipt of which a public office in the President's house is set aside; and, as this is a heavy tax upon the farmers—the principal producing class—and no proper accounts of expenditure are kept, a great unwillingness to pay the tithe is manifesting itself, and both pulpit and press continually call attention to the fact.

As is natural, from the peculiar tenets held by the Mormons, they have encountered much hostility. They were driven from Palmyra, in New York State, where the belief was first founded by Smith; when they migrated to Kirtland, in Ohio. Here, also, they were not long permitted to remain, and they then proceeded to the State of Illinois, settling near the town of Commerce, which they called Nauvoo, or City of Beauty. The country under their auspices soon changed its appearance, and the settlement became very thriving, until the animosity of the citizens of the State was aroused by certain acts of the new sect; when they rose in arms and expelled them, after having killed Joseph Smith. The Mormons, under the leadership of Brigham Young, now took refuge in their present home—the valley of the Salt Lake—which they have occupied for the past thirty years; but their stay there does not promise to be of much longer duration, as, since the completion of the Pacific Railroad, which brings them into communication with the Eastern and Pacific States, there has been a great irruption of "outer barbarians," attracted by the rich mining in the Territory. This contact with the Gentiles does not agree with the "peculiar institutions" of the Mormons, who now meditate a general exodus to Mexico, with the Government of which country President Young has negotiated for the settlement of a large tract of country, to be enrolled as a State in that Republic.

Polygamy, as has before been mentioned, was originally denounced by the Book of Mormon; and it is denied by many that Smith was the author of the Revelation of Celestial Marriage, but that it was added by Young and Pratt. Certain it is, that Smith's four sons do not give in their adherence to it, and have formed a schism in the church, having a large following under the name of "Josephites."

I went to the Tabernacle on Sunday to attend the service, which consisted of hymns, sung by a fair choir, with organ accompaniment; what seemed to be extempore prayers, delivered by several of the elders; a long sermon by the bishop, Orson Pratt, and the administering of the sacrament. The preacher deduced from texts,

taken from Daniel and Matthew, evidently to the entire
satisfaction of his hearers, that the world was near its
end, that the millennium was nigh, and that God's Kingdom—*i.e.*, the Mormons—was about to be saved. He
attempted to show that the signs of the times all tended
to point towards the second coming of the Messiah; and
further stated that, if asked what would replace the
various forms of government existing in the world, his
own opinion was, that the best parts of the constitutions
of the United States and Great Britain might be retained
in the coming Kingdom of God. The whole service, to
me, seemed wanting in solemnity, and the Bishop's address
sounded very much like rank blasphemy. The attendance at the Tabernacle could not have exceeded 2,500;
but I was assured that the paucity of the numbers was
in consequence of the bad weather, and that the general
average is very much larger. The appearance of the
congregation was poor in the extreme, and it would be a
difficult matter to find in any place of worship a more
unintellectual-looking lot of people.

And now a few words, before closing this chapter, as
to how the whole affair strikes a stranger. Any dispas-
sionate observer, after staying among these people a
short time, cannot fail but come to the conclusion that
the whole is a gigantic fraud; such an one, in fact, as
could only be originated and carried out by our American
cousins. The system enriches and aggrandizes the

leaders at the expense of the poor deluded beings—the rank and file—whose industry and indomitable perseverance have to bear the pressure of supporting the drones—their leaders—who live upon them. The greater part of the tithe undoubtedly finds its way into the pockets of Brigham Young, who subdivides among his leading elders according to desert,—*i.e.*, the extent of their subserviency to him. The various leaders of this peculiar people possess nice villa residences and drive in fine carriages, the President, Brigham Young, having no less than three large mansions, called respectively the Lion and the Beehive houses ; the third, recently erected, has not as yet received a name. As he has eighteen wives and forty-four children, it is easily imagined that much accommodation is needed. One of his wives, Ann Eliza Young, recently obtained a divorce from him, and published a book, exposing the evils of Mormonism, by which, it is said, she has realized a little competency. Though Brigham Young is the head of the Church, he yet takes a prominent part in lay matters. He is director of several companies, banks, &c., and is largely interested in a mercantile business called the "Zion's Co-operative Mercantile Institute." This latter is conducted on a very extensive scale, and has its ramifications and branches, not alone in the city itself, but throughout the whole Territory.

The rising generation, notably the female portion, are said to be averse to polygamy ; and, if such be the case,

this fact, combined with the increasing communication
with the outer world, may tend, in time, to stamp out
this great evil. At present, it is growing, and growing
rapidly, as, in addition to the excess of births over deaths,
there is a large immigration going in. Shortly after my
departure, an increase was made to their numbers by
the arrival of some five hundred immigrants under the
leadership of several elders.

CHAPTER V.

OVER THE ROCKY MOUNTAINS—CHICAGO.

DEVIL'S GATE—Weber Cañon—Devil's Slide—Echo Cañon—
Castle Rocks—Plains—Sherman—Prairie Dog Villages—
Omaha—Bridge over the Missouri—Burlington—History of
Chicago—Great Fire—Fine Position—Water Supply—Streets
—Parks—Public Buildings—Grain Trade—Cattle Trade—
Pork Packing—Hotels.

LEAVING Salt Lake City, I arrived at Ogden in time to catch the early train of the Union Pacific Railway Company, eastward bound.

From Ogden we passed through well-cultivated country, divided into farms, with comfortable-looking homesteads, and soon entered the Wahsatch Mountains through a chasm called the Devil's Gate, on a high trestle-bridge, elevated fifty feet above a torrent that dashed through the gorge. We were now in that region of grand and most imposing beauty called the Weber Cañon. Fortunate in passing through in the daytime, I had an opportunity of seeing this, the most interesting part of the route from the Pacific to the Atlantic. The road winds through the devious turns of this cañon, where rock-ribbed mountains—snow-capped—rise to an awful height on either side, destitute of vegetation, except here and there a stunted pine-tree. This rocky region lies between the valleys of the Salt Lake and the

Green River, and the train has to pass through five tunnels, having an aggregate length of 2,000 feet, cut through solid rock, which never crumbles, and does not require to be arched with brick.

Shortly after entering Weber Cañon we passed that wonderful natural rock formation called the Devil's Slide, which consists of two ridges of rock standing some ten feet out from the face of the mountain, up which they extend, parallel to one another, for a distance of, perhaps, 200 feet. For some thirty miles we continued our way through the dark deep cleft of this cañon, the rocks assuming most fantastic shapes, and the Weber River raging below. From the shape the rocks assume, they have received such names as Pulpit Rocks, Witch Rocks, &c.

Emerging from Weber Cañon, we soon again entered a defile in the mountains called Echo Cañon. This is a deep, rugged ravine, some seven miles in length, flanked on the left-hand side by bold precipitous cliffs, from 300 to 800 feet high, totally destitute of vegetation, and waterworn by the action of storms. The opposite side, sheltered from the southerly gales, is composed of sloping masses of rock, covered with moss and other vegetation; and of gently undulating hills. In the gully below, a beautifully clear stream flows placidly through the channel it has made for itself in the rock ; but about half-way down, the ravine narrows to a mere defile, and here the river seems to grow wilder, and, like an athlete, to gather up its strength to overcome at a bound the obstacles in its course. Here, too, the banks are steeper, the vegetation more luxuriant, and the lofty cliffs on the left are broken up into all manner of fantastic outlines. As we flitted rapidly by, it was easy to imagine that these masses of red rock, piled one above the other to such an enormous height, assumed the shape of any object uppermost in our minds ; but, on approaching the celebrated Castle Rocks, almost at the outlet of the cañon, it needed no stretch of imagination to picture a huge baronial castle, for there it seemingly stood before us, with its solid walls frowning down upon us, with its towers and keep, as if wrought by the hands of giants. Here are to be seen the massive boulders and huge masses of rock collected on the brink of the precipice, and intended to have been hurled down upon the enemies of Mormonism—the United States forces, under General Johnson, sent out against the Mormons in 1857.

After passing Echo Cañon, we arrived at Castle Rocks station, and shortly afterwards left the Territory of Utah, at Granger. The scenery continued of a mountainous character until we arrived at Green River station, where we began to traverse bleak and desolate plains, covered with that unpleasant alkaline dust which so effectually prevents the growth of anything but the hardy sage-bush. On these plains not a living thing is to be seen but jackass-rabbits and lizards, and this continues for two hundred miles. We soon re-commenced our up-hill progress, and could perceive how rapid was our ascent by the increased coldness and rarefaction of the atmosphere. Passing Laramie City, a rising little town that owes its origin and prosperity to the Pacific Railway, and where the Company have some machine shops, and, what is very praiseworthy, a good hospital for the cure of their employés when sick or in case of accidents, we soon crossed the Dale Creek bridge. This bridge is 650 feet in length, and through its interstices we could see the valley and little meandering stream 126 feet below us. It is built of timber, and has a very frail appearance, though capable of supporting the heaviest train.

Continuing our ascent, we soon reached Sherman, named after General Sherman, the "tallest" officer in the American army ; where we attained the highest point on the line—an elevation above the level of the sea of 8,250 feet. This is the highest station in the world, and we were literally above the clouds, for we could see them below us resting on the sides of the mountain. From this point we commenced to descend, through rugged granite hills, winding in and out of interminable snowsheds, until we arrived at Cheyenne City, the principal station on the line between Ogden and Omaha.

Cheyenne is already a place of some importance, with a population numbering about 4,000; and as it is the point of junction of the line to Denver, the capital of the State of Colorado, and the Pacific line, it seems likely to become a thriving and prosperous town. It is situated on a broad plain, watered by the Crow Creek, at an elevation of 6,041 feet, and is barely ten years old, the first house having been .built in 1867. It was at one time a very "rowdy" place, but its reckless times are now over, the worst portion of the population having been drafted off to other places. I may mention that we here obtained capital antelope steaks, which have a flavour between that of beef and venison.

Passing Cheyenne, we lost sight of the Rocky Mountains, that vast mountain chain that, with great variety of configuration and under different names, extends from the Arctic Ocean to the Straits of Magellan; that wondrous barrier which Nature seems to have reared to prevent the encroachment of the waters of the Pacific. Seen from this point, they look like white clouds in the distance.

For about two hundred and fifty miles we now passed through a vast grazing country, covered all the year round with a good nutritious grass. We saw several herds of antelope feeding on its verdurous slopes, and now and then caught a glimpse of buffaloes in the distance. A remarkable feature are the prairie-dog villages, where several thousands of these small animals live in communities, burrowing in the earth like rabbits. The Prairie-dog, or Wish-ton-Wish (*Spermophilus Ludo-*

vicianus) is a rodent; and its popular name is due to the short yelping sound which it utters, when alarmed, and which resembles the bark of a young puppy. It is a pretty little animal, measuring about sixteen inches in length. The head, being peculiarly flat, gives it a very remarkable appearance. It resembles the rabbit in many respects, burrowing, and, like it, is very prolific. As our train approached this " village," which occupies several hundred acres, honeycombed by a labyrinth of subterranean passages, we could hear the alarm given, and could see the little animals scampering off to their burrows, into which they disappeared with a comical little flourish of their hind legs. Their curiosity, however, seemed irrepressible, for presently we could see their little heads protruding cautiously from the burrows, and their inquisitive little eyes prying to discover the cause of the disturbance.

The prairies here appear boundless, stretching away as far as the eye can reach, and then disappearing into space. There are portions under cultivation, but these farms seem swallowed up in the immensity of the country. There is an undulating sweep, or "roll," in these prairies, that, combined with the want of trees, and the multitude of tiny blossoms on the turf, give them all the appearance of the ocean.

After passing the prairies we entered the valley of the Platte River, which, in this season of the year—the early summer, before the grasses and flowers have withered, and the streams still meander through it—is particularly attractive. The Platte River itself is broad, being fully three-quarters of a mile across ; but it is a sluggish stream, quite useless for navigation, as the water is said to be only some few inches deep.

We skirted its banks for some distance ; and, after passing several stations of more or less importance, arrived at Elkhorn, a pretty little place situated on the river of the same name, which is 300 miles long, and flows through a valley of good and productive land. This is quite a German settlement ; and as the settlers are at no loss for cheap and abundant supplies of food, it is a most thriving one. The river abounds with good fish, and game is abundant ; and each snug little farm-house seems to possess a good orchard and garden. The lines of these settlers have certainly fallen in pleasant places.

We now rapidly approached Omaha, the terminus of the Union Pacific and other lines; to which fact it owes its rise and progress. It is built on the western bank of the Missouri River, on a high ridge, which rises some fifty feet above the water level. Though barely twenty years old, it already possesses a population of about 25,000 inhabitants, and is a bustling, thriving place, with numerous fine buildings and two daily papers. It bids fair to become one of the most important of Western cities.

At Omaha we crossed the River Missouri, on the great

iron bridge, to Council Bluffs, on the opposite bank of the river. This bridge has only recently been completed, and replaces the old bridge of boats, which necessitated passengers descending from the cars and crossing the river on foot. This has been obviated by the completion of this great engineering work, which completes an unbroken line of railway from the Pacific to the Atlantic. This bridge is of enormous length, built entirely of iron, the abutments being hollow pillars, sunk below the bed of the river to a strata of rock, and filled with granite. ·Its cost of construction was £200,000. Electing to proceed to Chicago by the Burlington and Quincy Railroad, I took the cars of that company, and soon found that we began to move at a much quicker pace; for, whereas our progress from San Francisco had been only twenty-two miles an hour, we now proceeded at a rate of thirty-five. The road passed through fine farming country in the State of Illinois, many spots being perfect pictures of rural beauty, with here and there a town nestling in amongst the trees. Some of these towns are of great commercial importance. Burlington, a town of some 20,000 inhabitants, situated on the Mississippi • River, is the principal : Aurora and Galesburg are also large towns. At Burlington we crossed the Mississippi on a magnificent iron bridge.

After a continuous journey from Ogden of three days and a half, we arrived at Chicago. This incessant day and night travelling, would certainly be prejudicial to health ; were the jar and vibration of the bogie carriages, in use throughout America, as great as in the carriages on our own lines. Chicago is, in population, the third city of the Union; in commercial importance, it ranks after New York ; its population may be set down at 400,000. In 1830, it was only a trading station with the Indians, consisting at that time of a few log houses only ; and the city has attained its present proportions, since that comparatively recent date. What causes it, however, to be regarded as the most remarkable city in the world, is the fact, that since the great fire of 1871, the greatest conflagration of modern times—in which three and a half square miles of the principal part of the city were burned ; in which 17,450 buildings were destroyed, and 98,000 people rendered homeless ; the monetary loss of which, was estimated at little short of thirty-seven millions sterling, ten millions only of which, were covered by insurance—the city has sprung up like a phœnix out of its ashes, and now shows but few traces of the dire calamity. The buildings destroyed, have all been replaced by noble stone edifices; so that now Chicago may justly claim to be a " City of Palaces." In 1874, another fire broke out in the devoted city, and six hundred acres of buildings were consumed ; yet, in spite of these disastrous drawbacks, Chicago exhibits but few remains of either devastation.

The position of the city is most favourable, being
situated mid-way between the Eastern and Western
States ; on Lake Michigan, which gives it a splendid
water-way. It is also the centre of a vast railway system,
and has in consequence become the greatest grain
emporium in the world, and also the largest cattle and
lumber market. It has not been without straining every
nerve, that the city has been rebuilt in such an incal-
culably short time ; house rents are very high in
consequence, and ratepayers obtain some concession,
by paying their rates in advance.

Chicago is well laid out, the streets being generally
eighty feet wide ; some of them are from three to seven
miles in length, run due north to south, and east to west,
and cross one another at right angles. The Chicago
River runs through the city, and with its two branches,
divides it into three parts ; between which, communication is effected by means of thirty-three bridges. These
bridges swing on central pivots, and have to be opened
to allow vessels to pass ; but as this is found to interrupt
the street traffic, they are commencing to build tunnels,
under the bed of the river. Two have already been
completed ; one connects the northern and western
divisions of the city, and is 1,600 feet long; the other
joins the northern and southern, and is 1,900 feet in
length.

The water supply system is excellent. A tunnel has

been constructed, extending two miles under the lake: into this the water enters through a grated cylinder in an immense crib, on which a lighthouse, and signal-station have been erected. This tunnel connects with a tower 130 feet high, up which the water is forced, by four engines, having a daily pumping capacity of 72,000,000 gallons; and it flows thence to all parts of the city.

The principal streets are State-street (the Broadway of Chicago), Lake, Clark, La Salle, Randolph, Dearborn, and Maddison streets. These are principally devoted to business purposes; the best private residences are in Wabash and Michigan Avenues, which have a semi-suburban appearance, and are planted with double rows of fine trees.

The park system is unrivalled. There are six parks, covering altogether an area of 1,900 acres, which are connected by a series of fine boulevards, extending round the city, and forming thirty miles of fine drives; in addition to those in and around the parks. Lincoln Park is a good specimen of landscape gardening; it possesses fine trees, an artificial river, lake, and hills; and contains many summer houses, and rustic seats and bridges. Humboldt, Central, Douglas, South, and Union, are all nice parks; especially the latter, which only contains some seventeen acres, but is so well laid out, that it has the appearance of being much larger.

The principal public buildings were destroyed during

the fire, and whilst all, or nearly all the business premises have been rebuilt; the re-construction of the former, is being more slowly proceeded with. The new building to be used as a Custom-house and Post-office, now in course of erection, at an estimated cost of $4,000000, gives promise of being one of the finest in the country.

The Courthouse, the building of which is also being very slowly proceeded with, will be a fine edifice. The Chamber of Commerce is very elaborate in its interior decorations. A visitor, accustomed to the quiet and orderly conduct of English Chambers of Commerce, must be astonished, at the noise and excitement, that here prevail. It is devoted principally to the interests of the grain trade; and the samples of wheat, maize, flour, &c., are displayed on small tables, which are always surrounded by crowds of buyers and sellers, whose shouting and gesticulating, convert the hall into a perfect little pandemonium.

The grain trade of this city is colossal. In 1872, which be it remembered, was only one year after the fire; 86,000,000 bushels were received, and exported. The mode of conducting this trade is as follows: the railway brings the grain into the warehouses, which are immensely high buildings, situated on the river side. The grain is received, and shipped loose, the use of bags being dispensed with; it is shovelled into the elevators, and taken up to the top of the building, where it is run over to the other side, and discharged into the ships through big shoots. A day's transactions at one of the fifteen grain-elevating stores in the city, which have an aggregate storage capacity of 13,000,000 bushels; will give some idea of the magnitude of this trade. Some days 600 railway trucks are unloaded, the daily average being 370, each of which, contains 400 bushels; each elevator in the building, and there are twenty-four, raises 600 bushels per hour.

The cattle trade, the next largest industry of Chicago, is of almost equal magnitude. The value of the live stock imported in 1873, was $80,000,000. The trade is carried on outside the city, at the Union Cattle and Stock Yards, which are so extensive, that they merit a description. These yards comprise 345 acres, of which 100 are enclosed as pens. They have thirty-one miles of drainage, seven miles of roadway, 2,300 gates, and cost nearly £400,000 ; they have accommodation for 21,000 cattle, 75,000 hogs, 22,000 sheep, and 200 horses. Connected with the yards, are an hotel, a bank, a Board of Trade ; and a town of 4,000 inhabitants has sprung up just outside with churches, schools, &c.

Some of the pork-packing establishments are in close proximity to the yards. This is also a very extensive industry ; as in 1872, 1,500,000 hogs, and 16,000 cattle were packed. The process, seen for the first time, is rather interesting. The hog, pressed onward by those behind, proceeds up an inclined plane, through a door, which might appropriately be inscribed with Dante's

"All hope abandon, ye who enter here,"

into a pen, in the upper part of the packing-house. Here a chain, attached to a pulley in a sliding frame, is slipped over one leg ; and he is jerked up, his throat cut, his body lowered into a vat of boiling water, taken out, scraped, disembowelled, and hung up to cool. In this manner, a hog, which had ascended the inclined plane in all the pride of youth, strength, and porcine beauty ; was within a few minutes, lying in the lower story of the packing-house, cut up, salted, and ready for exportation.

The lumber trade is also one of the principal resources of the city, it having been computed, that in 1873, a billion feet of lumber were received.

Chicago possesses many fine hotels. The Palmer House at which I obtained excellent accommodation, though not the largest, contains 650 bedrooms, many with bathroom attached. The principal dining-room with its grand columns, and frescoes, is very beautiful ; and the drawing-rooms are splendidly furnished. Other large hotels are the Grand Pacific, the Tremont House, the Sherman House, and a host of others of secondary importance.

Whilst on the subject, I may mention, that American hotels offer great facilities to guests ; by having attached, post and telegraph offices, a bureau where may be obtained railway and theatre tickets, barber's shop, &c. The prevailing rates are from three to five dollars per diem, according to the accommodation required. There is no sociable table d' hôte; guests order from the bill of fare what they require, and it is brought up in portions, all at the same time, so that most of the dishes become cold, before they are partaken of. This arrangement seems to be, in consequence of a want of patience, on the part of Americans, to sit out a table d' hôte dinner; for they invariably appear to swallow down, as quickly as they can, the numerous dishes they order; and then to hurry off at once to their avocations.

Communication with all parts of the city, is obtained by means of a most perfect system of tram-cars, the fare being only six cents the course.

There are several fine theatres; McVickers' is one of the best in the country; the New Chicago is also a nice house, but the others are mediocre.

I left Chicago for Detroit early in the morning, and arrived there the evening of the same day.

CHAPTER VI.

DETROIT—THE NIAGARA FALLS.

DETROIT is a clean and pretty town, containing many fine villa residences; it is situated on the Detroit River—a noble stream, twenty miles long, connecting Erie Lake with Lake St. Clair; and forming here, the best harbour on the whole chain of lakes.

Detroit is the principal city of Michigan, and contains some 100,000 inhabitants; the river front for miles is lined with warehouses, dry-docks, shipbuilding yards, founderies, and grain-elevators ; and the city is laid out on the usual American rectangular plan.

The streets are broad ; those containing the business premises vary from 50 to 100 feet in width ; the avenues of private mansions from 100 to 200 feet. They are generally planted, with a double row of trees on both sides, and these beech, chestnut, oak, elm, and maple trees, all covered with foliage, give a pretty countrified appearance to the city, and form a most agreeable shade.

The Campus Martius, a fine open space or square, is the

principal feature in the arrangement of the city. Taking up the whole side of this square stands the City Hall, a noble stone building in the Italian style; consisting of three stories, above the basement, and surmounted by a Mansard roof. In the middle of the square is the Soldier's Monument; erected in memory of the Michigan soldiers who fell in the civil war. On the north side of the Campus Martius is the Opera House, one of the largest buildings of the kind, in the country. This square is crossed by Woodward and Michigan Avenues, and from it radiate Munroe Avenue and Fort-street. These avenues, for miles, are lined on both sides with detached, and semi-detached houses, each standing in its own garden.

In the centre of the city is a semicircular park, called the Grand Circus, which is divided in two by Woodward Avenue, each part containing a fine fountain.

The principal buildings are the Custom House, and Post Office, both under the same roof, and the Board of Trade. The Great Wheat Elevator of the Michigan Central Railway Company is a large building, from the cupola of which a grand view is obtained of the city, river, and Lakes St. Clair and Erie.

The manufactures of the city are rather extensive, comprising ironworks, machine-shops for the construction of railway rolling-stock; flour mills, breweries, and extensive tobacco and cigar factories; the shipping

interests are also large. Pork and fish-packing are carried
on very extensively.

In the river are the pretty Belle Isle, Grove Isle, and
Put In Islands. The former is a pleasant resort for pic-
nic parties : on Grove Isle are many beautiful summer
residences, and Put In Islands are largely visited by
excursionists.

On leaving Detroit, the train ran on to a large steam-
ferry, and it was so conveyed over to the Canadian side
of the river ; whence the route lay through a fine agricultural country, with some charming bits of scenery, in the
Province of Ontario.

We passed several flourishing little towns ; the rising
city of London being amongst the number, and finally
arrived at our destination, Clifton, a village on the
Canadian side of the celebrated Niagara Falls.

What can I say of this stupendous work of Nature,
that has not been said before ?—how describe the awe
and feeling of insignificance, that overcome one, at the
first sight of this "Thunder of Waters?" It is futile to
attempt to give an idea, of the awful grandeur of the
various scenes about the great cataract ; and no pen
could accomplish an adequate description of them.

So much has been written about the Falls of Niagara,
that their general features are now well known to most
readers ; still a few words about them may not be out of
place.

The first European who saw Niagara was a French missionary, Father Hennipen, in 1678, nearly two centuries ago ; and since that time, it has been visited by millions of people, who have all in turn gazed with feelings of awe, on this tumultuous rush of waters. Niagara is an old Iroquois word signifying Thunder of Waters, and a more expressive one could not have been found; it being in very truth a thunder of waters, the noise of which, is heard at a distance of many miles, sounding like the moaning of the sea.

All the great lakes of America, viz., Superior, Huron, Michigan, and Erie, pour their waters into Lake Ontario, through a channel about thirty-six miles in length, called the Niagara River, which forms part of the boundary between Canada, and the State of New York. The famous Falls are about twenty miles below Lake Erie, and are divided in two by Goat Island, and called respectively the American, and Canadian, or Horseshoe Falls. The former is 900 feet wide and 164 feet high ; the latter 2,000 feet wide, and 158 feet high. Over these precipices, the irresistible water rushes, at the rate of one hundred millions of tons per hour ; and this immense volume of water is computed to wear away the rock, by friction, at the rate of a foot a year ; and the Falls are said to have gradually receded from Queenstown, seven miles below, to their present position. The river varies in width, from half a mile to three miles ; at the Falls it is three-quarters of a mile wide, and is studded above, with numerous islands. The total descent from Erie to Ontario is 334 feet, which, distributed equally over the length of the river, would be a fall of ten feet in the mile,—or hardly sufficient to form a rapid.

Just above the Falls, a magnificent cast-iron bridge has been built across the river ; and standing upon it, one forms some idea of the grandeur of the scene ; in seeing the river, as it whirls along to its impending fate in a rushing torrent ; seeming as if it would carry the bridge, with its puny passengers, over the frightful precipice. This is called the Rapid above the Falls ; the descent being sixty feet in the mile. In its course the river foams, and hurls up its billows, as in a very ecstacy of madness ; and forms a fine contrast, to the grand and placid flow of the waters over the Falls.

One of the best views of the great Cataract is obtained from Prospect Park, on the American side, which consists of some eight acres of wooded land, skirting the river bank, for some distance above, and below the American Fall. It is well laid out in lawn, and walks ; and the bank in the immediate vicinity of the Fall has been built up with solid masonry, a low wall protecting the brink ; so that visitors can look down, in perfect safety, into the dizzy depths of the precipice on the verge of which they stand. This stone parapet is projected to the very edge of the Fall, and one can stand just by the rush of the waters, and take in at one view, the whole magnificent arc of the Niagara.

The vivid hues of the waters, as they glide onward, and bend in an unbroken sheet over the brink, contrast with the whiteness of the wreathing mists, into which they plunge. The ceaseless, and stupendous movement of the descending deluge ; the huge rising clouds of spray ; the deafening roar, that arises from the boiling depths ; all combine to bewilder and confound one. Only after long gazing is the mind able to realize, that it confronts a picture, the beauty and sublimity of which, can never be wholly grasped.

A better conception of the majesty of the great Cataract can be formed, by viewing it from the huge rock-masses, that lie in chaotic confusion about the foot of the American Fall. The mountain of descending waters impends almost directly above the spectator ; the boiling abyss is close at his feet. A breath of air will suddenly turn upon him a blinding shower of spray, through which, if the sun be shining, he may catch a view of the iris ; not as elsewhere, a prismatic arc only, but an almost perfect circle of dazzling radiance.

To obtain the best idea, however, of the astounding magnitude of Niagara, and to take in its immensity ; it is necessary for a visitor to put on an oilskin dress, obtained on payment of a small fee, and, accompanied by a guide, to go under the Fall. After recovering the fatigue of the descent, and getting the spray out of his eyes ; he will feel amply repaid by the scene that bursts upon his astonished gaze. I availed myself of this convenience, and stood behind the watery curtain, twenty feet thick in the centre, under the great Horseshoe Fall ; with the water falling over the ledge, right in front of me, and descending to the river in one gigantic bound. The view here is awfully grand. As we gaze upwards at the frowning cliff, that seems tottering to its fall, and pass under the thick curtain of water,—so near, that it seems as if we could touch it; and hear the hissing spray ; and are stunned by the deafening roar, that issues from the misty vortex at our feet ; an indescribable feeling of awe steals over us, and we feel like very insects, before this manifestation of one of Nature's greatest terrestrial wonders. Behind our narrow footpath, the precipice of the Horseshoe Fall rises perpendicularly to a height of ninety feet ; at our feet, the cliff descends about seventy feet, into a turmoil of foam ; whilst in front of us is the liquid curtain, which, ever passing onward, never unveils this wildest of Nature's caverns.

Goat Island is half a mile long, by a quarter broad ; it contains about seventy acres, and is very heavily wooded. It is situated on the very verge of the Falls, and appears as if it would be hurled into the depths below, by the force of the water. From Goat

Island, some of the best views are obtained, and it is connected by bridges with Bath and Lunar Islands. Even this glorious scenery is not sacred from the desecrating hand of utilitarianism ; for on Bath Island a large paper mill has been erected ; thus utilizing the enormous water-power, for industrial purposes.

From Lunar Island, the best view is obtained of the beautiful lunar bow, which is only visible for a short time in the month, when the moon is full, and high in the heavens. The solar bow is always seen, when the sun shines on the Falls. It is said that Lunar Island trembles, and this does not seem at all improbable, although I did not observe it.

Three lovely little islands, called the Three Sisters, lie on the American side, in close proximity to Goat Island ; and being now all connected by bridges, are easily accessible, and from them is obtained, the grandest view of the rapids.

The Cave of the Winds is another and perhaps one of the most remarkable sights here ; it has been formed by the action of the water, wearing away the softer strata of the rock, and thus making a cavity 100 feet wide, 130 feet high, and 30 feet deep. Along the floor of this cave, the spray is dashed with considerable force, striking the walls and curling upwards along the roof ; thus causing the turmoil, which has procured for it, its name. When the sun shines, a beautiful rainbow, quite circular

in form, quivers amid the driving spray.   It is necessary
to wear the oilskin dress, when visiting this cave ; and
all conversation is impossible, the mighty Fall claiming its
right of alone being heard.

. The Suspension Bridge, which spans the river in view
of the Falls, is a splendid structure 1,240 feet long; it
will support a weight of 3,000 tons, and itself weighs only
250 tons, or only one-twelfth of its sustaining power.
Three thousand people might be distributed over its
length, without in any way affecting its supporting
capacity.   In the tower on the Canadian side, is an
elevator, which takes visitors to the top ; whence a fine
view of the surrounding country is obtained.

The Whirlpool Rapids, about two miles below the
Falls, are interesting ; for here the waters rush along at
the rate of twenty-seven miles an hour, tossing up breakers
to the height of thirty feet.   According to Sir Charles
Lyell, fifteen hundred million cubic feet of water rush
through this gorge every minute.   To see these rapids
from the best point of view, it is necessary to descend to
the water's edge, by means of a hydraulic lift constructed
for the purpose, on the American side of the river.

About three miles below the Falls, the river takes an
abrupt turn, and the water dashes with great violence
against the cliffs on the Canadian side ; forming a large
whirlpool, in which, the waters seem to seethe, and boil,
as though in a caldron.

The Clifton House is a comfortable hotel, in the immediate vicinity of the ground, upon which was fought the
Battle of Lundy's Lane.

After having revelled in these glorious scenes, and
stamped them indelibly in my mind, never to be effaced
until such time as all will be oblivion ; but to be recalled
by memory in years to come, to prove a fruitful source
of pleasurable reminiscence ; I reluctantly tore myself
away from this place, so rich in Nature's grand sights.
The train soon conveyed me to Lewiston, a little town
situated at the junction of the Niagara River and Lake
Ontario ; formerly of some importance, but of late much
injured by the opening of the Erie and Welland Canal.

At Lewiston, I embarked on a steamer proceeding to
Toronto, where I arrived in a couple of hours' time, after
a pleasant trip across Lake Ontario.

CHAPTER VII.

TORONTO AND TRIP DOWN THE ST. LAWRENCE.

DESCRIPTION of Toronto — University — Queen's Park — Public Buildings — Sunday Observance—Difference between the People of Canada and the United States—Kingston—Emperor of Brazil—The Thousand Islands—Timber Rafts—Rapids of Long Sault—Lake St. Francis—Cedar Rapids—La Chine—Canals.

TORONTO is, after Montreal, the largest city of Canada ; it has a population of 60,000 ; and is situated on a beautiful bay, the entrance to which is narrow, and protected by a long spit of land, called Gibraltar Point. The city has a very English appearance, and the streets, though not wide, are generally well paved. The principal are King and Yonge streets, which contain all the best retail establishments. Front street, containing large blocks of warehouses, is also a fine thoroughfare.

The University of Toronto is a grand edifice, in the pure Norman style of architecture ; the buildings forming three sides of a quadrangle. It is a noble institution, and one of which Canadians may justly be proud ; it possesses a library of 20,000 volumes, and a fine museum of natural history. Adjoining the University is the Queen's Park, comprising fifty acres of well laid out ground ; approached by long avenues of chestnut,

beech, elm, and oak trees. In the park is a fine monument, surmounted by a large figure of Britannia; erected
to the memory of those Canadians, who fell in repelling
the Fenian invasion of 1866.

The Post Office, built in the Ionic style, is a fine
building; the City Hall is plain and unpretentious; the
Custom House and Court House are handsome structures. Osgoode Hall is a very imposing, and extensive
building of the Ionic Order; it contains the Provincial
Courts, and has a fine law library. Amongst other
buildings of a public character may be enumerated
St. Lawrence Hall, the Masonic Hall, and the Exchange.

Toronto is to Canada, what Edinburgh is to Scotland,
the seat of its higher educational system; and contains
many fine public schools, and colleges. Knox College,
a Presbyterian institution, is one of the principal, and
most successful. Trinity College is a spacious building,
with numerous turrets and quaint gables; and is situated
in the midst of extensive grounds. The Normal School,
for the training of teachers, the Model Schools, and the
Educational Museum are plain buildings; but standing
together, as they do, in a large park, they have a very
picturesque appearance.

The Provincial Lunatic Asylum, and the General
Hospital are well-conducted institutions; in the latter a
large number of patients are annually cured. Toronto

contains more public institutions than any city in the United States, double, or even treble its size; and the same may be said equally of Montreal, and Quebec.

The churches are very fine, notably the Episcopal and Roman Catholic Cathedrals; both being good specimens of ecclesiastical architecture.

Sunday is kept in a very puritanical manner; the hotel bars being closed from seven o'clock on Saturday evening until six o'clock on Monday morning; and no vehicles are to be seen in the streets.

I was much struck with the difference in manners, and appearance, between the people of the Dominion, and those of the United States. Though descending from the same stock, and though only separated by a river, the difference is very perceptible, and the longer I stayed in Canada, the more it impressed me. People there seem quieter, less addicted to extravagance in language, and dress; more polite and self-denying; slower-going certainly, but more reliable.

All commodities are cheaper in Canada, than in the States; in consequence of the baneful effects of the pernicious protective system, into which the latter drifted; the fruits of which, are now being reaped in the high price of all articles entering into general consumption; in the great decrease of the mercantile marine; and in the general stagnation of trade, at present existing.

Proceeding on board the steamer *Spartan*, we had a pleasant trip on Lake Ontario, which large inland sea, 240 miles in length, we traversed all night; and arrived next morning at six o'clock, at Kingston, and entered upon the glorious scenery of the St. Lawrence.

At Kingston, the Emperor and Empress of Brazil, with their suite, came on board, travelling *incognito*. On leaving my cabin, early in the morning, I came upon an old gentleman, dressed in a rather seedy suit of clothes, and wearing a somewhat dilapidated slouched hat ; who at once entered into conversation, by asking me, in broken English, some question about the locality. When he heard that I came from Australia, he became most eager in his inquiries about its prospects, resources, and rate of progress. This gentleman, I afterwards learned, was the Emperor, who continued the whole day to mix unreservedly with the passengers ; on occasions, elbowing his way through the crowd in a very bourgeois manner. Royalty, I conclude, on close inspection, is not very different to ordinary humanity.

After leaving Kingston we found ourselves entering amidst that wonderful and beautiful collection of islands, known as the "Thousand Islands," in the lake of that name. These lovely islands commence a little beyond Kingston, and extend to Brockville, a distance of nearly fifty miles, in one continuation of lovely scenery ; forming the most numerous collection of river islands in the world.

They consist of about eighteen hundred isles, and
islets, of every conceivable form and size ; some being
mere bits of rock; others several acres in extent, thickly
wooded, and covered with beautiful vegetation.   At
times our steamer passed so close to these islands, that a
pebble might have been thrown on shore ; whilst
looking ahead, it appeared as though further progress
were barred; until on rounding the next point,
amid winding passages, the way would gradually open
before us.

Again the river would seem to come to an abrupt
termination, but on approaching the bank ; a channel
would suddenly appear, and we would be whirled into a
magnificent lake, bounded apparently by an immense
green bank, which would, on our approach, be moved
as in a kaleidoscope, and a hundred small islets, appear
in its place.

A feature of the St. Lawrence navigation, are the large
rafts of timber met on the river, which are floated down
to Montreal, and Quebec.   They are very large and
require many men to navigate them, and from the
number of huts erected on them, they have all the
appearance of floating villages.

The first place of any consequence we came to was
Clayton, a pretty little town, on the American side of
the river.   Alexander Bay, the next place passed, is
surrounded by a massive pile of rocks, and its situation

is romantic, and picturesque, in the extreme. We now left the Lake of the Thousand Islands, and re-entering the St. Lawrence, soon arrived at Brockville, a thriving Canadian town of about 7,000 inhabitants. Opposite to it, is the American town of Ogdensburg, a busy place, with a population of some 9,000 souls. The towns on the American side appear to have a greater appearance of prosperity, than their rivals on the Canadian bank.

After passing Prescott, which has a decayed appearance ; the increased rapidity of the current, indicated that we were approaching the first of that series of remarkable, and celebrated Rapids of the St. Lawrence ; and preparations were made to shoot it. A tiller was attached to the rudder, and manned ; whilst the Indian pilot took his station at the wheel, on the upper deck.

The Rapid of Long Sault, so called from its great length, rushes along at the rate of twenty miles an hour. When our steamer arrived within its influence, steam was shut off, and we were carried onwards by the force of the stream alone ; the water presenting all the appearance of an angry sea. The boat strained, and laboured ; its motion—quite different to the ordinary pitching and tossing—causing this going down hill by water, to prove a very novel sensation, and one not unalloyed with fear. Great precision is requisite, in steering a vessel over these rapids, her head having to be kept straight

with the course of the stream ; for if she diverge in the
least, she "broaches to," and is instantly capsized.

Shortly after leaving the Rapid of Long Sault, the
course of the river is entirely through British Territory.
We passed two small places known respectively as
Dickinson's Landing, and St. Regis ; and entered an
expansion of the St. Lawrence called Lake St. Francis,
which is forty miles long, and contains a number of
islands.   Leaving this lake behind us, we called in at a
small place, called Coteau du Lac, or St. Ignace ;
remarkable for its quaint old buildings, which give it all
the appearance of an old Norman village.

We now approached the Cedar Rapids, and the
passage through them was particularly exciting ; a
peculiar motion being felt, as the steamer glided from
one ledge to another, as though she were bumping on
the rocks, and settling down.

In close proximity were the Cascade Rapids, after
shooting which, we entered another expansion of the
river, called Lake St. Louis, and finally arrived off the
village of La Chine, adjacent to the Rapids of that name,
the most celebrated of the series.   The name was given
to it by the early settlers who thought they had discovered
the route to China.

Around all these rapids large canals have been
constructed, which prevent the upward navigation of the
river from being impeded.   The length of these canals

altogether is forty-one miles, with twenty-seven locks, and the cost of construction was enormous ; they reflect great credit on the energy of the people.

After safely passing the La Chine Rapids, we rapidly approached Montreal, and as we passed under the great Victoria Bridge, the view of the city, with the sun setting behind Mount Royal, was magnificent.

CHAPTER VIII.

MONTREAL, QUEBEC, AND OTTAWA.

MONTREAL, the commercial metropolis of Canada, is a nice clean city of 110,000 inhabitants, containing fine blocks of buildings ; but especially notable, for the number, and beauty of its churches. It is situated on the Island of Montreal, which is thirty miles long, and ten broad, and is formed by the junction of the Ottawa, with the St. Lawrence. It is connected with the main land, by the splendid Victoria Bridge, called by Montrealers " the eighth wonder of the world." This celebrated bridge is a mile and a half long, or with its approaches, nearly two miles ; is supported upon twenty-four piers, and two abutments of solid masonry ; the span between the centre piers being thirty-three feet. The iron tube, through which the lines are laid, is twenty-two feet high, and sixteen feet wide. This bridge was erected by Robert Stephenson, and though rather

an eyesore, obstructing as it does, the view of the noble
river ; is the only means of traffic with the mainland,
when the river being frozen over, navigation is impeded.
The city is laid out very much like an old French town,
but has of late been much improved; the streets are
gradually being widened ; and the city with its fine
blocks of buildings, and its numerous public edifices,
looks substantial, and shows the evidences of great
accumulated wealth.

The Quays are very fine; built entirely of solid limestone, and extending for several miles, undisfigured by
unsightly warehouses, and stores, as is unfortunately so
often the case.

At a short distance from the city is the Mont Real,
which gives to the city its name of Montreal, now used in
place of the ancient Indian one Hochelaga. From the
summit of this beautiful eminence, a magnificent panorama
is presented of the city, with its noble quays, and glittering spires. From it can also be seen, the great St.
Lawrence, as it winds its way, for miles, through a gently
undulating plain, dotted with pretty little villages. This
" mountain," as it is locally called, is a general recreation
ground for the people of Montreal ; and round the base,
and winding around the hill, are beautiful drives.

St. James, and Notre Dame streets, are the principal
promenades, and contain the best retail establishments.
St. Paul street, fronting the quays, and extending their

entire length, is the principal commercial thoroughfare;
whilst other smaller streets, branching off these main
arteries, contain large blocks of warehouses. The
principal residences are in the suburbs, extending to
the foot of the mountain.

Foremost amongst the public buildings is the Bonsecours Market, an imposing stone edifice, surmounted by
a large dome, which forms a very prominent landmark; in
the upper story is a large hall capable of accommodating
four thousand people. The Custom House is a large
building with a fine tower; the Post Office in St. James
street is simple in style, but very handsome; the Court
House is also a fine edifice of the Ionic order, and
contains a law library of six thousand volumes. The
Banking Corporations have very imposing edifices, notably
the Bank of Montreal and the Merchants' Bank. Other
prominent buildings are the City Bank, the Bank of
Ontario, the Young Men's Christian Association, the
Mechanics' Institute, and the Merchants' Exchange;
these are all ornaments to the city, and would do credit
to London, or New York.

The principal educational institution is M'Gill College,
beautifully situated at the foot of the Mount, and
possessing a fine museum. The Catholics, the dominant
denomination here, possess some grand edifices for
educational and other purposes. The Seminary of St.
Sulpice is a massive pile of stone buildings, of enormous

size; without any claim to architectural beauty, but more resembling an immense barracks. The Grey Nunnery is another vast building, designed in form of a cross. The Black Nunnery, and the Convent of the Holy Name of Mary, are both very extensive. The Hôtel Dieu, for the cure of the sick, is large and imposing, and is, together with St. Patrick's Hospital, under the charge of the Sisters of St. Joseph.

To enumerate the churches would require a volume; they are all fine, and many of them are very elaborate structures; but in going through the city a stranger is puzzled to imagine where all the funds came from, to erect all these costly edifices. Certainly a very large portion of the property in, and around Montreal, is owned by the Catholic priesthood. The Romanist Cathedral of Notre Dame, in the Place d'Armes, off St. James-street, is the largest place of worship on the continent; seating from ten to twelve thousand people. It is built of stone, in the Gothic style, and has six towers, the two front ones being each 213 feet high. Its interior decorations are tawdry, and in very bad taste; it possesses however, a fine peal of bells, the largest of which, the Gros Bourdon, is said to weigh thirteen tons. Even this large structure will be exceeded in size, by the new cathedral in course of erection, on the plan of St. Peter's in Rome. The Anglican Cathedral, though small, is a perfect specimen of English-Gothic architecture. Amongst the most note-

worthy of the Episcopal Churches are Christ Church
Cathedral, Trinity, St. George's, St. Thomas', and St.
Stephen's. The principal Catholic Churches are Bishop's
Church, St. Patrick's, which seats five thousand people,
and has very fine stained glass windows ; the Church of
Gésu, which has the finest interior in the city ; Recollect,
Bonsecours, and St. Mary's. The Nunneries and Semi-
naries have also chapels attached. The leading Presby-
terian place of worship is St. Andrew's. The Unitarians
possess a fine building called the Church of the Messiah.
In addition to the above is a legion of other Churches,
of all denominations.

Though Montreal is 600 miles from the sea, its
commerce is large ; owing to its advantageous position
at the head of navigation of the St. Lawrence, and of the
great chain of improved inland water, which extends
from the Lachine Canal, to Lake Superior. It is the
principal shipping port of the dominion ; its imports in
1873 amounted to nine millions sterling, and its exports
to five. In addition to this large shipping trade, Montreal is the seat of numerous industries ; the principal
being the manufacture of agricultural implements, axes,
saws and tools, steam engines, boots and shoes, paper
and furniture, etc.

I left Montreal for Quebec, by the steamer City of
Montreal, in the evening ; and arrived at the latter city
early the next morning. These river steamers are most

commodious, and beautifully fitted up ; and form, in the warm weather, a far more agreeable mode of travelling, than the dusty railway cars.

The first appearance of Quebec from the river is very grand ; here the banks rise precipitously, and are covered with fine trees, nestling amongst which are numerous private residences. The river widens considerably, and is commanded by the guns of the citadel, perched high up on the height ; from which the ramparts extend round the city, rendering it so strong a fortress, that it has been called the " Gibraltar of America." Since the regular troops who garrisoned it were recalled, and the charge of the fortress assumed by the Canadian Government ; the ramparts are falling into decay, and the guns rusting. The British Government commenced the building of three powerful forts on Point Levi, opposite to Quebec, on the other side of the river ; but left them in an incomplete state, and nothing has been done by the Canadian Government to finish them. It would consequently not be a difficult matter, for an enterprising enemy, to carry Quebec, as it stands, by a *coup de main ;* and this fortress, formerly considered impregnable, in the hands of an enemy commanding the navigation of the St. Lawrence, would virtually place Canada at his feet.

It seems a pity, that the Canadian Government is so apathetic in this respect ; and the day may come, when it will bitterly rue it. The ramparts enclose a circuit

of three miles, within which is the city proper, which with
its two suburbs of St. Louis, and St. Johns, is called the
Upper Town. The Lower Town is built round the base
of the promontory. The city is situated at the junction
of the rivers St. Charles and St. Lawrence; and has a
population estimated at 60,000; it is the oldest town in
Canada, and ranks in importance after Montreal. The
streets of Quebec are very narrow, and steep; especially
those connecting the Upper, with the Lower Town. A
large portion of the Upper Town is taken up by the
buildings of the great religious corporations of the
dominant denomination; such as the Seminary, Laval
University, the Ursulines, the Hôtel Dieu, and the ancient
Jesuit College; which latter has more recently been used
as a barracks.

Over the remaining portion of Upper Town, not taken
up by the fortifications, are crowded the quaint, mediæval
streets, and houses; the latter built generally of stone,
two storeys high, and roofed with a shining tin, which
gives the city, when the sun shines on it, a very glittering
appearance. The Lower Town is the oldest, and most
important; here are clustered at the foot of the promontory, under the guns of the grand battery, 200 feet above,
the principal wharves, ferry-landings, commercial houses,
and banks.

The principal shipbuilding yards, a large industry here,
are situated on the banks of the St. Charles; whilst the

coves of the St. Lawrence are covered for miles, with vast rafts of timber ; the export of which, constitutes the chief trade of Quebec. Durham Terrace is an esplanade in Upper Town, on the very brink of the precipice ; it is the great promenade of the inhabitants, and from it, the view is magnificent.

Just before I arrived, a large and disastrous fire had taken place in Montcalm Ward, in which four hundred houses were destroyed ; the appearance of the burnt district was very saddening.

On the opposite side of the St. Lawrence, are situated the populous towns of South Quebec, New Liverpool, and Point Levi.

The public buildings of Quebec are very unpretentious ; the churches alone having any claim to architectural beauty. The Marine Hospital, built after the style of the Temple of the Muses, on the banks of the Ilissus, is one of the few public buildings worthy of notice ; as is also the Laval University, an offshoot of the Catholic seminary, an imposing building, or series of buildings ; in connection with which are a fine laboratory, geological, mineralogical, and botanical collections, a museum of zoology, and a library of some fifty thousand books.

The Market Place in the Upper Town, presents a very interesting appearance, when the peasants, or as they are called, "habitans," bring in their produce for sale.

The drives about Quebec are very beautiful ; all along the St. Louis and St. Foy roads, are fine residences and gardens.

The Plains of Abraham, memorable for the great victory of General Wolfe, which gave Britain an American Empire ; and in the course of which both General Wolfe, the British, and General Montcalm, the French Commander fell ; are being gradually encroached upon, by the suburbs of St. Louis and St. John ; but enough still remains, to mark the battle field. A modest column has been here erected, by the British army, to the memory of Wolfe, on the spot where he fell ; and an obelisk stands in the town, in honour of Wolfe and Montcalm jointly.

Near Quebec, are the celebrated Falls of Montmorency, which are wondrously beautiful. The River Montmorency here falls over a ledge of rock 250 feet high, in a volume of snow-white foam, fifty feet wide. The heights on either side are well wooded; and with Quebec, and the St. Lawrence in sight, form a scene of most surpassing beauty. Near the Falls, are the Natural Steps,—a series of ledges, cut in the rock, by the action of the water ; each step being about a foot high, and as regular as if wrought by human hands.

Before leaving Quebec, I crossed over to Point Levi ; and a pleasant drive brought me to another natural wonder, the Fall of the Chaudière. Over a mass of rock, 150 feet high, the water falls in a sheet 350 feet wide ; presenting all the appearance of boiling water, whence its name Chaudière or Caldron.

Returning to Montreal by the steamer *City of Quebec*, in the day time, I had an opportunity of seeing the scenery of the St. Lawrence, on this portion of its long course. For some distance from Quebec, the high, well-wooded banks give the river a very picturesque appearance ; but approaching Montreal, the country becomes flat, and uninteresting.

I now took the steamer up the River Ottawa, to visit the city of that name,—the political metropolis of the Dominion. This stream flows through thickly-wooded country, with extensive saw-mills on the banks, for the conversion of pine logs into planks, etc.

Ottawa is situated on the river of that name, at its junction with the Rideau ; the city being on a hill, between the two rivers, and divided into the Upper and Lower Town, by the Rideau Canal. It has a population of 25,000 ; is the centre of the lumber trade of the Ottawa and its tributaries ; and otherwise possesses some extensive flour-mills, and manufactories of agricultural implements, mill machinery, &c. Facing it, on the opposite side of the river, are the suburbs of Hull, and New Edinburgh, connected with the city by several bridges.

The great feature, however, of Ottawa, is that imposing pile, or rather piles of building,—forming as they do

three sides of a quadrangle,—occupied as the Parliament
Houses and Government Offices.   These buildings are
situated on a hill, descending abruptly to the river ; are
built in the Italian Gothic style ; and, with their irregular
towers, and pinnacles glittering in the sun, form as
imposing a structure, as is to be seen in America.   The
south side of the quadrangle is formed by the Houses of
Parliament ; whilst the east and west sides, are taken
up by the Departmental Offices.   The two wings
containing the latter, are quite detached from the front,
forming the Parliament Houses ; and there are thus three
distinct blocks of building, which together form a pile, of
which Canadians may justly be proud ; and a fit abode,
for the Legislature of the great colony.

The interior of the Houses of Parliament is very fine ;
the Chambers for the two branches of the Legislature have
beautiful stained glass windows, but are still very light.
The Chamber of the Commons, to my mind, is far
handsomer than its prototype at Westminster ; but like
it, its acoustic properties are bad.   Facing the Speaker's
chair is a fine statue of Her Majesty, and busts of the
Prince and Princess of Wales ; and the corridors contain
portraits of former Speakers.   The Parliament Library
is a handsome structure, containing forty thousand
volumes.

The most important edifice in the city, after the
Government buildings, is the Catholic Cathedral of

Notre Dame. This is a large stone building, with two spires covered with shining tin,—poor in style,—but having a very picturesque appearance, when seen from a distance.

The scenery in the vicinity is very beautiful; consisting of well-timbered country, watered by several rivers, interspersed with numerous waterfalls. The principal of these are the Falls of the Chaudiére, and the Rideau; the former being especially beautiful.

A curious feature of Ottawa, is the turpentiney odour of the atmosphere; caused by the number of saw-mills. The river from the same cause is covered with a layer of saw-dust.

On my return to Montreal, I took the train, across the Victoria Bridge, for Plattsburg,—a small town on Lake Champlain, in the State of New York.

Before leaving Canadian territory, I may mention, that I was disappointed with the slow growth of the country; which is attributable in a great measure doubtless, to the severity of the winters, during which time, the country is ice-bound; Halifax being the only port on the Atlantic seaboard, open all the year round. Still Canada, with its fine soil; its comparatively close proximity to the European markets, and consequent immigration advantages; and with its great rivers and lakes, ought to be very much in advance of Australia—a country of so very much more recent colonization. This is, however, by no means

the case ; Canada being behind Australia in every
respect, except farming and manufacturing. To my
mind, a comparison between the people of the United
States, and Canada ; or between Australians, and Canadians ; is greatly to the prejudice of the latter. There
seems to be a want of energy, in the Canadians, which is
attributed mainly to the large French element. With
their fine water power, cheap carriage, and moderate cost
of labour, manufactures should be more numerous ; but
the fact is, the excessive number of wealthy religious
corporations absorbs a great portion of the available
capital of the country ; which would otherwise be employed, in developing its vast, and as yet barely touched,
sources of national wealth.

Canada has magnificent resources in its immense tracts
of fertile country, boundless forests, and great mineral
wealth ; what is wanted, is population to develop them.
The new Province of Manitoba is alone, supposed to be
able to support a population of fifty millions ; and yet
comparatively little is done to tap the continuous flow of
emigration to the States. It cannot be too often iterated,
that the want of Canada is the same as that of Australia,
viz., small capitalists, skilled mechanics, and practical
agriculturists. For such men there is a fine field in those
great Colonies ; and to these, my advice is Emigrate.

With the increasing difficulty experienced, in finding
employment for young men, of the middle class ; and in

view of the number, that are earning a precarious livelihood as clerks in merchants' offices with small prospect
of promotion ; but with all the cares, and heartburnings,
incidental to the struggle to keep up a respectable appearance, on insufficient means ; would it not be better, to
give them a special training, that would enable them to
make careers for themselves in the colonies? Repeating
what I have said above ; the classes of men most likely
to do well in the colonies, are skilled agriculturists, and
mechanics, who can at all times earn good wages ; until
they accumulate sufficient capital, to start on their own
account. Now, the young men who are sent out from
England have, as a rule, received a general education,
that fits them only for town life ; and they are fortunate if
they obtain situations as clerks; as the colonies are
glutted with young men of this class, who without capital,
have little chance of making positions for themselves.

Would it not therefore be desirable, in training young
men for colonial life, to do away with the common idea,
that manual labour is incompatible with the feelings, and
status of a gentleman ? In new countries especially,
where the line of demarcation between the different
classes of society, is neither so fine drawn, nor so marked ;
labour should be considered ennobling, and a gentleman
by education, be no less a gentleman, because he prefers
to guide the plough, rather than keep a set of books in a
merchant's office.

Would education unfit a man for the drudgery of manual labour? Or, would it rather be a means of giving him, a decided advantage over his less-educated fellow-workers? I imagine, the majority of thinking people will incline to the latter belief; and come to the conclusion, that a good education would be an incentive to sobriety, and perseverance; with the assistance of which, a man would soon push his way in the world; to say nothing of the many modes of rational amusement, and means of embellishing his life, that education would open up to him.

If it be granted, that it is desirable to train young men specially for colonial life; the question then arises, how such training could best be attained? And I reply, by the formation of good schools of agriculture and technology. Instead of sending them to an ordinary school, where they receive a general education; let it be made easy for fathers, to send their sons to schools, where in addition to general learning, they would be taught a trade; and then, after serving for a time on a farm, or in a workshop, where they would learn practically what had, at school, been taught theoretically; they would then become useful additions to young countries, where land can be obtained on easy terms, and the mechanical trades are well remunerated. With such a general and technical education, and the help of a little capital, they could, in a few years' time, make for themselves very comfortable positions.

There is a large class of young able-bodied men, whose abilities are not of the highest order; and who cannot settle down to the drudgery of office routine. For these especially, a good technical education would be of the greatest advantage; and their superabundant energy, which in many cases, unfortunately, leads them into the dissipation of the large cities, could be rendered useful, and profitable, to themselves and others, in pastoral and agricultural pursuits.

I have been induced to make these remarks, from having seen so many fine young fellows, who have gone out to the colonies with every wish to work; but not having been able to obtain employment as clerks, the only occupation their education fitted them for; they have squandered their small means, and eventually found themselves a burden on their friends.

CHAPTER IX.

GENERAL REMARKS ON CANADA AND THE COLONIAL QUESTION.

PROVINCES of the Dominion—Government—Population—Imports and Exports—Treatment of Indians—Loyalty of the People —Withdrawal of the Troops—Ignorance in England on Colonial Affairs—Independence of the Colonies *versus* Consolidation of the Empire—Objects of a Customs' Union.

THE Dominion of Canada comprises the various provinces of Quebec, Ontario, New Brunswick, Nova Scotia, British Columbia, and Manitoba, formerly known as the Red River Settlement. Newfoundland and Prince Edward's Island, will doubtless also shortly enter the Confederacy.

Each Province has its local legislature, which generally consists of two Houses, and a Lieutenant-Governor; except in the case of Ontario, where only one Chamber exists. They also return to the General Parliament, which meets at Ottawa; Senators to the Upper, and Representatives to the Lower House. Senators are appointed by the Crown, for life, and have the title of Honourable. Members of the Executive have seats in either House; and are dependent upon the support of a majority. The Governor-General is appointed by the Crown.

Canada Proper consists of the Provinces of Ontario, and Quebec, formerly called Upper, and Lower Canada. The latter is inhabited chiefly by French Canadians, a steady, industrious, but very conservative race ; possessing none of the go-a-head spirit of British Colonists, and much influenced by their spiritual heads. When the Province was ceded to Britain by treaty, in 1763, its laws and religion were guaranteed. This was very well, so long as the community continued small ; but now the anomaly exists of French law prevailing in the Province of Quebec, which contains the populous Cities of Montreal, and Quebec ; and English law, in the adjoining Province of Ontario. This gives rise to endless litigation, and the laws will no doubt be, in time, assimilated by the Federal Government.

The population of the Dominion is at present about 4,000,000, it having in spite of the small immigration, increased five hundred per cent. within the last half century. The 40,000 French subjects, ceded to Britain by the treaty 1763, have by excess of births, over deaths, and aided by a small immigration from France, increased since then to 1,000,000 ; which is scarcely to be wondered at, when the large families of the "habitans" are taken into account.

Forty-five per cent. of the whole population is Catholic ; the Church of England, the Presbyterians, and the Methodists, together count a similar per-centage ;

whilst the balance is made up of the smaller denominations.

The exports of Canada amount to £15,000,000; and
the imports to £24,000,000. It is curious, that whilst
57 per cent. of the imports come from Britain, and 34
per cent. from the United States; the exports are exactly
in the inverse ratio, for the States take 57 per cent., and
Britain 34 per cent. of her products and manufactures.

The Mercantile Marine comprises 7,000 vessels of all
sizes, with an aggregate tonnage of over a million tons,
and an estimated value of seven millions sterling.

The progress of the colonies, now forming the
Dominion, in earlier times more especially, was assisted
in a great measure by their security from attack by the
Indians; consequent on the system adopted towards
them by the colonists, viz. of making *bona fide* bargains
with them for their lands; paying them by annuities,
held in trust for them by the Colonial Governments;
and leaving them certain "reserves," to be theirs in
perpetuity. This system contrasts favourably, with that
in force in the United States; and the history of the
Dominion is not tarnished, like that of the States, by
accounts of risings of the Indians, to retaliate upon their
spoilers, for the wrongs inflicted upon them.

The sentiment of loyalty to the Sovereign, and
institutions of the Mother country, as in Australia, is
very general; and it seems a pity that so little is done

in England, to show that the feeling is recognized, and appreciated. On the contrary the tone of the English press generally, is to show the colonies, that they are regarded as a source of weakness; and it is only the well-known fact, that this opinion is not shared by the bulk of the British people, that holds the colonists to their allegiance.

It is not my purpose here, to enter into the question of whether the colonies are, or are not, a source of weakness; but it is apparent, that so long as fresh fields are requisite for the surplus population of Britain; it is far better, that the energy and productive power, thus lost to that country, should go to build up the remoter portions of the Empire, that may in time be able to materially assist the Mother country, in an emergency; than that they should assist in developing the resources of rising young nations, rendered by coldness, and neglect, inimical to Britain, and that might in time, come to regard its downfall, solely in the light of a dangerous competitor removed out of the way.

It was a mistaken policy of the Home Government to withdraw the troops from Canada and Australia; but more especially so, in the case of the former, situated as it is, in such close proximity to a power whose lust for new territory is proverbial. Apart from the fact that these troops served as a visible proof of Britain's intention to protect her colonies in time of war, and

9

as a model and standard of precision and regularity, to the militia and volunteers; but they were also the means of diffusing much useful information about the colonies, and were to colonists a constant reminder of their connection with the Mother country.

It is a frequent complaint, that much gross ignorance prevails in Britain on colonial affairs; and it is unquestionable, that were more prominence given in schools to colonial geography, and other means used, to disseminate information concerning the requirements, resources, and prospects of the Colonies, a large portion of the flow of emigration to the United States might be diverted.

There is no living man, who has such reason to be proud of his birthright as an Englishman. When he goes to America, and sees that great offshoot of Britain, the United States, and that rising nation, Canada; when he thinks of the great future of Australia, New Zealand, and the South African colonies; and remembers our great Indian Empire; he must be dead to every patriotic feeling, if the knowledge that he belongs to that wonderful Anglo-Saxon race, that has effected such great things, does not cause him a sensation of pride. And yet, there is a section of politicians in England—let us hope it is a small one—who openly advocate—for the purpose of taking a penny off the income tax—the voluntary giving up of all this. Is the proud boast of Britons, that they possess an empire on which the sun never sets, to be weighed by a pounds, shillings, and pence standard? Would it not be better, if those politicians, who advocate the casting adrift of the colonies, were to devote their thoughts to a proper consolidation of the Empire; by means of which, the resources of every portion might be united for mutual interests and defence.

Deprive Britain of her colonies, and you reduce her to the status of Holland; consolidate, and an Empire would be formed, in comparison with which, that of ancient Rome would sink into insignificance.

The time is coming when this question will be brought home to politicians; and a decision will have to be arrived at, whether to draw more closely the bonds that unite the colonies with Britain, or to altogether sever them. Certainly the time is not yet ripe for an Imperial Council, to legislate on matters affecting the Empire at large, as has been of late so often suggested; but a step in that direction might well be taken, by the formation of a Customs' Union between its component members; by means of which, their various products and manufactures could be mutually exchanged, without any of those vexatious hindrances that at present exist.

This would be a means of knitting more closely together, the different parts of the Empire; and as the Dominion of Canada, the Australian and South African colonies, increased in population and wealth, then the system could be extended, to enable them to have a voice in matters affecting their general interests and common weal.

Bearing in mind the rapid, yet steady progress, of the Colonies ; the time cannot be very far distant, when a Confederation, with Britain as the centre, will be deemed advisable ; and it ought to be a source of pride to Britons, to think, what a power for good such a Confederation would be : how it would spread over the world the British Laws, Language, and Institutions ; and being unaggressive, and devoted to the Arts of Peace ; what a high destiny it might attain ; and how materially it might assist in bringing about, that concord between the Peoples of the World, that should exist.

CHAPTER X.

LAKES GEORGE AND CHAMPLAIN, SARATOGA, TROY, AND ALBANY.

PLATTSBURG—Lake Scenery—Fort Ticonderoga—Rogers' Slide— Narrows—Caldwell—Neighbourhood—Glen's Falls—Saratoga Hotels—Springs—Life at Saratoga—Saratoga Lake—City of Troy—Position of Albany—Description of the City—Green-bush.

A SHORT journey brought me to Plattsburg, a place of some importance, on Lake Champlain, in New York State; where I took the steamer, to proceed down Lakes Champlain and George; one of the most beautiful, and enjoyable trips, in America. Lake Champlain is 120 miles long, being very much larger than Lake George, and though abounding in glorious views, does not present such lovely bits of scenery, as the latter. After leaving Plattsburg, a fine panoramic view presented itself to our gaze; blending in rare beauty, the wildest mountain scenery, with placid water-views. The lake expands, as far as the eye can reach; and its waters are broken by countless islands, and headlands. The important town of Burlington reposes in calm beauty; and beyond are seen the Green Mountains, their summits clearly defined upon the horizon. On the other side, Lion Mountain, and the numerous peaks of the Adirondacks are visible.

A short distance down the Lake, rises a lofty peak of the Green Mountains, called the Léon Couchant, which by a slight stretch of imagination, takes the form of a lion with head, mane, and paws complete. Here also are situated four small islands, called the Four Brothers, lying almost exactly at the four cardinal points. A delightful trip through these various beautiful scenes brought us to Fort Ticonderoga, a ruin standing on a high rocky eminence. This fort was built by the French in 1756, and was named by them Carillon ; but this has been supplanted by its present Indian appellation of Ticonderoga, which signifies "noisy," so called from the Falls in the vicinity. Though now a ruin, it has been the scene of many a fierce struggle in former times. In 1758, it was attacked by an English force of 16,000 men, under Abercrombie, who was however repulsed, with a loss of 2,000 men ; but in the following year, it was abandoned by the French, and taken possession of by the British. Ticonderoga has been held successively by the French, the British, the French again, the American Colonists, and was finally reduced by General Burgoyne ; when it remained in the hands of the British, until the termination of the revolutionary war, and was then ceded to the Americans.

The scenery in the vicinity of this old fortress is very beautiful, and the drive of four miles along the sparkling stream, that connects Lake Champlain with Lake

George, is delightful. This turbulent little stream, has a descent of 230 feet; in the course of which, there are two series of lovely cascades, called the Falls of Ticonderoga. The romantic beauty of these, is however, marred by the factories that have been erected along the banks, near the villages of Alexandria and Ticonderoga. The scenery however, varies continually; and openings in the foliage, reveal vistas of the lake, with the hills and valleys of Vermont in the background; whilst the little stream foams, and tumbles, beside us, presenting at every turn, new and beautiful combinations, of rock draped with rich verdure, the colours of which harmonize beautifully with the dark blue of the lake, glimpses of which we are continually obtaining.

We here embarked on board the steamer *Minehaha*, to proceed down Lake George. This lovely lake, only thirty-six miles long, is at an elevation of 320 feet above the sea-level, and is not much wider than a large river. It unites in its scenery, the soft and gentle, with the grand and magnificent. It has been appropriately compared with Loch Katrine, and is in fact the Trossachs on a larger, and grander scale. The water of this lake is quite clear and pellucid; and the bottom is composed of a fine yellow sand, which is visible at a great depth. The banks are surrounded by high hills, covered with verdure; and the waters are studded with numerous charming islets. It seems a pity that the native name of

Horicon, signifying Silvery Waters, has not been retained; for it is most appropriate, especially when the moonbeams are reflected on the rippling waters.

Proceeding on our way down the lake, we soon passed two bold promontories, standing on either shore, called respectively St. Anthony's Nose, and Rogers' Slide. The latter derived its name from a ruse, practised upon the Indians, by Major Rogers, who was escaping from them; by which he persuaded them, that he had actually slid down the declivity, which is some 500 feet deep, with a precipitous front of naked rock. It happened thus :—The Major was escaping from the Indians, on snow shoes, during the winter; and eluded pursuit, until he reached the brink of this tremendous cliff. Aware that they would follow his track, he descended to the top of the smooth rock; and casting his knapsack, and haversack of provisions, down upon the ice of the lake ; slipped off his snow shoes, and, without moving them, turned himself about, and put them on his feet again. He then retreated along the southern brow of the rock, and made his way safely down a ravine to the lake below ; where he snatched up his pack, and made his escape, on the ice, to Fort George. The Indians in the meanwhile coming up to the spot, saw two tracks, both apparently approaching the precipice ; and concluded, that two persons had cast themselves down the rock, rather than fall into their hands. Just then, they saw the bold leader of the Rangers making his way across the ice; and believing, that he had slid down the steep face of the rock, considered him under the special protection of the Great Spirit, and made no attempt at pursuit.

The next place of interest that we came to was Sabbath Day Point, so called, because here, General Abercrombie, while on his way to attack Ticonderoga, landed on the Sunday, to rest and refresh his army. The scenery here is most lovely; excelled only by that we now commenced to pass through,—that portion of the lake called the Narrows. The hills at this point, extend into the lake, and contract it considerably; while their height render this contraction more impressive. The Black Mountain rears its bulky form here, to a height of about 2,200 feet; and around, are the boldest, and most picturesque parts of the shores of Lake George. The water though 400 feet deep, is so pellucid, that the eye can penetrate far down into its mysterious depths. The lake is here studded with numerous beautiful islands; and in passing through this lovely scenery, view follows view, like the moving picture of a panorama; filling the eye with ever-changing visions of beauty, and raising expectation to its utmost pitch, as headland after headland is passed, and the various lovely scenes are gradually unfolded. The beauties of the Narrows of Lake George are quite beyond the power of language to describe.

For some time we steamed through scenes of ever-changing grandeur and beauty, past islands and islets, of all sizes and forms, some of which are of great extent, level and cultivated; others rise in rugged cliffs, from the water, their summits crowned with tufts of vegetation; others again are mere dots, rising but a few feet out of the water; but all are lovely, and interesting, to the tourist, who has the good fortune to visit Lake George.

We now rapidly approached the head waters of the lake, and after calling in at numerous small places, all picturesquely situated; we were landed at Caldwell, a remarkably pretty place, situated on the margin of the lake, at the foot of high hills.   I at once proceeded to the fine Fort William Henry Hotel, from the grand piazza of which, the view is most exquisite.   There are many fine drives through the beautiful scenery about Caldwell; and hither, during the summer months, flock numbers of the wealthier Americans, many of whom possess fine summer residences on the shores of the lake, and in the hills.   The whole of the district is famous for its historical associations; here on the placid waters, or along the shore have the pioneers of civilization and the Indian inhabitants met in deadly strife, con-tending for its possession; here, have the British and French forces, encountered each other, in fierce feud; and here finally, have the British, and American colonists met, to fight out their differences.

There are many ruins of forts, and places memorable for some celebrated encounter, or romantic adventure ; and the interest of the spectator, in these various lovely spots, is enhanced by the associations connected with them.

After spending a couple of days, amid these romantic scenes—time not permitting a longer stay—the stage-coach conveyed me, over a fine plank-road, and through beautiful scenery, to Glen's Falls,—a rising town on the Upper Hudson. Here the river rushes over a ledge of rock, 900 feet wide ; and falls a distance of 70 feet, presenting a very picturesque appearance. It is, however, difficult to imagine this, to be the spot, where Cooper laid most of the scenes in his " Last of the Mohicans ;" for civilization with its usual disregard for sentiment, and its all-conquering utilitarianism ; has covered the banks with saw-mills, and other useful, but decidedly unsightly buildings.

The Rensselaer and Saratoga Railroad rapidly conveyed me to the celebrated watering-place, Saratoga, where all the upper ten of American society flock during the months of July, and August, to drink the waters, to dance, and flirt ; for in Saratoga, in these consist the whole duty of man, and more especially so, of woman. Although at the time of my arrival, the great influx of visitors had not yet set in, it was no difficult matter to form an idea how gay the place must be, in the height of the season.

Saratoga contains a population of 9,000 ; but during the months of July and August, this number is increased by visitors to 30,000. The hotels are colossal ; the principal being the Grand Union, the Congress Hall, and the United States, which are only open for four months out of the year, and will alone repay a visit to Saratoga. The Grand Union has over a mile in length of piazzas, two miles of halls, ten acres of carpets, and eight hundred bedrooms. These hotels are, in spite of their great size, very comfortable ; and their proprietors cater for the amusement of their guests, during the season, by having good bands, and balls every night. They will each accommodate from a thousand to fifteen hundred guests. The arrivals at Saratoga, during the season, often amount to a thousand daily.

And now a few words about the Springs, which are chalybeate or acidulous saline, according to the relative proportion of their particles ; the constituent ones being carbonate of soda, chloride of sodium, carbonate of magnesia, hydriodate of soda, silica and alumina, carbonic acid gas, with occasional traces of iodine and potassa. The principal are Congress and Columbia Springs, both situated in a tastefully laid out little park, and over them elegant light structures have been erected ; Excelsior Spring, situated in Excelsior Park ; High Rock Spring, situated on the top of a rock, whence its name ; the Hamilton and Hathorn Springs, in the centre of the town; the Washington, United States, and Empire Springs; Red Spring, so called from the colour of its waters when agitated; the Star, Eureka, and White Sulphur Springs. Putnam Spring is used for bathing purposes, being chalybeate. The most remarkable is the Geyser, or Spouting Spring, the waters of which spout up high in the air, through a shaft sunk 140 feet through the solid rock; and are very saline, and only fourteen degrees above zero. The Glacier Spring is another geyser, in close proximity to it.

Life in Saratoga, in the great hotels, may be summed up in a few words. It is ephemeral, only lasting a couple of months; and during that period wealth, beauty, fashion, and other ingredients not so desirable, meet and intermingle, in the whirl and excitement of the ballroom at night; visits to the Springs in the morning; and promenades or drives, in the afternoon. The extravagance of the present fashion of ladies' dress, is here seen in all its ungraceful, and disfiguring effects. Here, really pretty girls of eighteen or twenty, may be noticed, painted, and powdered, and with their figures so distorted; that instead of natural youth, and comeliness, they present an appearance, more closely approaching decrepitude, and ugliness.

One of the principal jaunts in the neighbourhood, is the fine six mile drive to Saratoga Lake, a charming spot, where during the season, many regattas take place.

Taking train again, a pleasant journey through some of the fine midland counties of the Empire State,—as the State of New York is called, brought me to the important city of Troy.

This remarkably pretty town contains many fine public edifices, churches, and private residences; and has a population of 45,000. It is the seat of some very extensive manufactures, the principal of which are large iron-works, and factories for the manufacture of carriages, boots and shoes, and hosiery.

A notable feature, and a prominent landmark, is the Theological Seminary of St. Joseph, a fine structure of the Byzantine order of architecture, on Mount Ida. The city also contains the great Watervliet Arsenal, where are kept many relics of the battles that were fought around Saratoga and Yorkton. Six miles from Troy is the city of Albany, the capital of the Empire State, built on the River Hudson, at the head of its tide-water.

Albany, after Jamestown in Virginia, is the oldest settlement in the original thirteen States. It was founded by the Dutch in 1612, and named by them Williamstadt, until 1664, when it fell into the hands of the British, and was re-named by them Albany, in honour of the Duke of York and Albany, afterwards James II.

The city has a population of 80,000, and is a place of great commercial importance, from its position on the

Hudson, and its proximity to the great Erie and Champlain Canals; the former of which gives it the command of a fine water-way to the west; and the latter, facilities of cheap water carriage to the north. It is also the centre of a large railway system.

Albany is situated on a high ridge, and with its numerous spires, presents a very imposing appearance, when seen from the opposite side of the river. State-street, its principal thoroughfare, ascends from the water's edge to the height, on which stand the Capitol, and principal public buildings.

The present State House is a plain, unpretentious building; but the new Capitol now in course of erection, in close proximity to it, will be, when completed, the finest edifice in the whole country; with the exception alone of the Federal Capitol at Washington.

The State Library is a fine building, containing 90,000 volumes; and amongst other historical relics and curiosities, the original Arnold and André Correspondence.

The City Hall is a beautiful white marble structure, with a fine portico, supported upon six Ionic columns.

Albany possesses several public libraries, museums, and establishments devoted to educational purposes; amongst the latter may be mentioned the Normal School, for the training of teachers. Its churches are numerous, and many of them are handsome structures.

In the western part of the city is Washington Park, which forms an agreeable recreation ground ; and bids fair in time, to become a great ornament to it.

On the opposite side of the Hudson, and connected with the City proper, by means of a fine iron bridge, is the pretty and populous suburb of Greenbush.

CHAPTER XI.

THE HUDSON RIVER, WEST POINT.

SCENERY of the Hudson—Poughkeepsie—Devil's Danskammer—
Newburg Bay—West Point—Military College—Nomination of
Cadets—Training—Discipline—Cozzens—Buttermilk Falls—
Gibraltar — Sing-Sing —Sunnysides — The Palisades—Arrival
at New York.

AT Albany, I embarked on one of the magnificent
Hudson steamers, for the trip down the river to New
York. The scenery of the Hudson has been so often
described, that its beauties have now a world-wide
reputation, and attract crowds of tourists. It is often
compared with the Rhine, and by many thought to
exceed that river in grandeur; but I cannot concur in
this view. There are spots of rare beauty, such as
West Point, which favourably compare with any portion
of the Rhine scenery; but the greater width of the
Hudson, does not compensate, for its deficiency in those
romantic hills crowned with castellated ruins, which are
so numerous on the banks of its rival, and which
constitute one of its chief attractions. Still it is a grand
river, and the trip from Albany to New York, is most
enjoyable.

From Albany to Hudson, a rising little town of some
13,000 inhabitants, situated on the east bank of the

river, forty miles distant from the former, there is a degree
of sameness in the scenery ; but thence to Poughkeepsie,
the banks are lined with fine country residences, their
lawns sloping down to the water's edge. Poughkeepsie
lies nearly mid-way between Albany and New York; being
distant from the former place eighty miles, and from the
latter seventy-five. It is beautifully situated on hills overlooking the river, and was originally a Dutch settlement.
It has now become a fine town with a population of over
20,000 souls, containing several fine educational establishments, the best known being Vassar College for females.

After leaving Poughkeepsie, the scenery became much
finer ; and for some time we journeyed through pretty
country, passing many important, and pleasantly situated
villages. The most striking feature, on this part of the
route, is a broad flat platform of rock, jutting out into
the river, called the Devil's Danskammer, or dancing
chamber ; used until recently by the Indians, in the
performance of their religious rites.

We now entered the broad expanse of Newburg Bay,
and soon arrived at a small place called Cornwall, much
visited by tourists in summer time, being cool, near West
Point, and having many lovely drives in the neighbourhood. We pursued our way across the Bay, until we
came to its outlet ; with the Boterberg Mountains rising
on the one side, and Breakneck Rock on the other ; and
with mountains, and cliffs surrounding us on all sides,

seeming to shut us in the basin of the lake, lying at their feet, from which there appeared to be no outlet.

The principal of these mountains is Cro'-Nest, which rises abruptly from the water's edge, to a height of 1,500 feet; and being covered with verdure, presents a charming appearance. This is one of the most picturesque spots on the Hudson. Continuing on our course, past overhanging cliffs, with their background of hills; we soon came to West Point, which, from its beautiful position, and historical associations, is undoubtedly the most interesting of all the many charming places on the river.

In order to see the famous Military College here, I landed, and spent a most delightful day, amongst the great natural beauties of West Point. The College buildings are of stone, built on a platform of rock 150 feet above the river; and are approached by a fine road cut through the solid rock.

The view from this platform is splendid: cliffs rise abruptly from, and high hills undulate to the water's edge; and the river itself is covered with the white sails of numerous craft; whilst in the distance are seen the Catskill Mountains.

The College buildings comprise a large and somewhat imposing barracks for the accommodation of the cadets; a chapel and library, a mess-room, an observatory, laboratory, and riding-school. In front of the College is a fine piece of ground, on one side of which, are several pretty villas inhabited by the professors ; and the ground itself descends through woods to the river.   Here on the slope, and along the banks of the river, many pretty winding walks have been made, one being denominated "Flirtation Walk" ; and these certainly form a charming adjunct to the place.   In various spots, adding to the general picturesqueness, are erected batteries, and trenches, for the instruction of the cadets.

The nominations to the College are made by Members of the House of Representatives, and the course of study extends over four years, during which time a cadet is taught a knowledge of all branches of the profession ; so that, if he be recommended for a commission, and appointed say to a cavalry regiment ; his knowledge of infantry tactics, of artillery practice, and of engineering, may stand him in good stead in after life.   The value of this training was shown in the late civil war, during the course of which, nearly all the men who came to the front, were graduates of West Point.

The cadets wear a grey uniform ; and the discipline maintained is strict, in fact, it seems to be unnecessarily severe.   For the first two years after their admission, cadets are not allowed to go outside the College limits ; they are then permitted to visit their friends for two months, after which vacation they return for another term of two years.   The use of beer, wine, spirits, and tobacco, is prohibited ; and to prevent the possibility of smuggling the latter, cadets were forbidden the use of pockets in their uniforms. This stringent regulation has of late been modified, as it was found to act detrimentally; inasmuch, as after a forced abstinence from indulgences, that might otherwise have had little charm for him, a graduate, on leaving College, and finding himself placed beyond this forced restriction, usually rushed into the opposite extreme of over-indulgence. The cadets number 250, and the entire cost of study is defrayed by the State.

Reluctantly I left West Point, and again taking the day steamer, proceeded on my way down the river to New York. A short distance from West Point is a fashionable place of summer resort called Cozzens, where a fine hotel is erected, which during the hot months, is crowded with visitors. This hotel is perched on the top of a high cliff, the highest for a great distance around; and nothing can be more picturesque than its position high up in the air, looking down on the noble river. It is several hundred feet above the water; but so perpendicular are the rocks, that it appears double the height. The view of this building crowning the beetling cliff, which rises abruptly out of water, may well be compared with the most beautiful Rhine scenery.

Near Cozzens are the Buttermilk Falls, formed by a small river rushing down the side of a hill, a distance of 100 feet, and falling into the river below, in a sheet of white foam.

Another spot of unequalled grandeur is at a place

called Gibraltar : here the river makes a sudden bend to the west, and the Dunderberg Mountain lifts its towering head, just behind the prettily situated 'village. Close by is the great rock promontory of Anthony's Nose, rising 1,300 feet out of the water, and forming one of the most noticeable features of the river scenery. Opposite, on the other side of the river, protected by two strong forts, a creek empties itself into the river, of such depth that the largest ships could ride at anchor in it ; and in close proximity is the beautiful Iona Island, forming altogether as charming a view as could anywhere be seen.

We had now passed through the highlands, and the scenery, which for the last sixteen miles had been truly grand, now became more tame. Sing-Sing was soon reached : here is erected the great State Prison of New York State, which is visible from the river, is built about 200 feet above its level, and accommodates one thousand inmates. On the opposite bank of the river is Rockville, notable for its ice stores : here immense quantities of ice are cut out of a lake of fresh water in the neighbour-hood, and stored for the supply of New York City. The consumption of ice in this country is enormous. As we steamed past, being a hot day, it was very refreshing to see the great blocks of ice pushed down inclined planes, into lighters specially constructed for their conveyance. Hard by is the Croton Lake, which supplies New York with water, conducted through an aqueduct thirty-three miles long.

The private residences on the heights now became more numerous, and though many of the marble and stone buildings are more pretentious, none are so pleasing as the charming cottage called Sunnysides, formerly the residence of Washington Irving, and the place where most of his works were written.

Before approaching New York we passed that most remarkable, but singularly beautiful rock formation called the Palisades. These Palisades form an unbroken line for fifteen miles, of bold perpendicular trap rock, columnar in formation, from three to six hundred feet high ; which presents a solid front to the river. This rocky barrier is so effective, that the Hackensack River runs parallel with the Hudson, but at a higher level, for thirty miles, and at a distance of only two to three hundred yards.

We now approached Manhattan Island, upon which is built the City of New York, and as we steamed down the Spuyten Duyvel,—a branch of the Hudson,—separating New York from its dependency, Jersey City ; the change from the quietude of the scenery through which we had been passing, and in which I had for some time been lingering ; to the crowded shores, and busy waters ; where the noisy hum of active life resounded, and where the very atmosphere seemed to grow thicker, and more oppressive over this human hive ; I involuntarily sighed for the quiet of the rural scenes, through which we had recently passed.

CHAPTER XII.

THE City of New York, the commercial metropolis of the
United States, and the largest city in the Western
Hemisphere, is situated at the mouth of the Hudson, and
occupies the whole extent of Manhattan Island; which
is thirteen and a half miles, long, by a width varying from
a few hundred yards to two miles and a quarter. It also
takes up a portion of the mainland, and three small
islands in East River—viz., Randall's, Ward's, and Black-
well's. The area of New York is forty-one and a half
square miles, of which twenty-one and a half are on Manhattan Island, nineteen on the mainland, and about three
quarters of a mile on the smaller islands. Its greatest
length is sixteen miles, and its greatest width four and a
half. It is separated from the City of Brooklyn by a
branch of the Hudson called East River; and from its

other leading dependency, Jersey City, by the Hudson proper. The older portion of the city is irregularly laid out ; the larger and newer portion is more regular. The streets running the length of the island are called avenues, and are numbered from First upwards ; the cross-thoroughfares bear the name of streets, and are numbered from First and go up to 225th. The buildings between two streets form a block, and twenty-one blocks, including the roadway, measure a mile.

I was much struck with the English appearance of the city, and subsequently observed this to be a feature of all the great cities of the Eastern States ; notably of Boston, Baltimore, and Philadelphia.

Places like Chicago and San Francisco, being of much more recent settlement, have an individuality of their own ; but the large towns of the New England and other Eastern States, being substantially built, and having in parts already an old appearance, greatly resemble large English cities. In New York, especially, the streets near the wharves, in the older portion of the city, are very dirty, and have already acquired a somewhat antiquated look.

The Harbour is remarkably fine, and picturesque. The outer bar is at Sandy Hook, distant eighteen miles from the Battery at the extreme end of Manhattan Island ; and is crossed by two channels, either of which will admit vessels of the heaviest draught. Entering the Bay from the ocean, the Narrows are passed ; and vessels then sail between Staten, and Long Islands ; passing which, the batteries of Fort Richmond, and Fort Tompkins, are seen on the one side ; and Fort Hamilton, and Fort Lafayette, on the other. The vessel then enters the Harbour ; the whole of the city being spread out like a panorama in front, with Brooklyn on the right, and Jersey City on the left.

The population of New York now consists of one million souls ; while Brooklyn contains nearly half-a-million more. It is computed, that on every working day, there are a million and a half of people congregated in the city.

Broadway, the principal thoroughfare, is eighty feet wide ; it is six miles long, and contains most of the best retail establishments, the number and size of which are very great. The leading hotels with a few exceptions are also situated in this splendid street ; and it is a pre-vailing habit, to sit on chairs placed on the footpath, in front of them, greatly to the inconvenience of passers-by. This selfish attention to individual comfort, and total disregard of that of others, seems to be, by the way, very common with the Eastern Americans. So far, in fact, is this habit carried, that it becomes a want of common politeness ; for it is a daily occurrence, to see a lady enter a car filled with men and youths, none of whom, would think it necessary to rise, in order to give her a seat.

Wall Street, running from Broadway to the East River, about half-a-mile in length, is the financial centre of the City ; it contains most of the Banks, the Custom House, and Treasury.

The Bowery, the principal street in the eastern part of New York, is a continuation of Third Avenue; and is a fine, wide, and busy thoroughfare.

Fifth Avenue is the Belgravia of New York, almost exclusively devoted to private residences, of the better class ; and also containing some of the finest places of worship in the city. These residences very much resemble the second-class houses, in the West-end of London ; but have a greater number of steps in front called " the stoop," upon which, in the hot weather, the whole family may be seen sitting, in the cool of the evening. On Sundays, after Morning Service, numbers of well-dressed people congregate in this avenue, to promenade, and exhibit their elaborate toilets ; and it then presents a gay, and animated scene.

Maddison, and Lexington Avenues, are scarcely inferior to Fifth Avenue, in the number of their private residences. Park Avenue is a very broad street, with fine houses ;. and for some distance is tunnelled, to allow the street cars to pass underneath ; and where this occurs, the openings for admitting light, and air, to the tunnel, are surrounded by little greens in the centre of the avenue, which give it a unique, and very pretty appearance.

Large as the city is, it boasts but few squares, which would contribute so much to its health, and beauty, and serve to relieve the monotony, of the miles of bricks and mortar; and the few it possesses, have a very untidy look. Maddison Square on Broadway, is the most fashionable of these reserves; it covers six and a half acres of tastefully laid out ground, and contains a fine monument to General Worth, the hero of the Mexican War. Union Square, also situated on Broadway, is an oval of three and a half acres, well laid out, with fine trees; and contains a fine fountain in the centre, a bronze equestrian statue of Washington, and a monument to Lincoln.

Mount Morris Square, on Fifth Avenue, comprises twenty acres, with a rocky eminence in the centre; round which a walk winds to the summit. The Battery, at the south extremity of the city, looking out upon the Bay, embraces twenty-one acres, laid out in walks, &c., and protected by a granite sea-wall. It was the site of a fort in the early days of New York, and later was surrounded by the residences of the wealthy; now, this portion of the city in which it is situated is wholly devoted to business purposes. In close proximity used to be the well-known building called Castle Gardens, which has recently been destroyed by fire. This was for a long time past, the place where immigrants were lodged on arrival; to prevent their falling into the hands of the many land-sharks, who would otherwise have plundered them of their little all.

Washington Square covers eight acres ; Stuyvesant Square four and a half acres; and with Gramercy Park, and Reservoir Park, complete the number of New York Squares ; which are quite inadequate to the requirements of the city.

Every succeeding year more clearly demonstrates the utility of squares in densely populated cities, to act as breathing places, or metaphorically speaking, as lungs.

Communication with the different parts of the city is effected by means of a good tram-car system ; rails being laid through most of the streets, except Broadway, where omnibusses take the place of the tram-cars. There is also an elevated railway, raised above the streets, so as not to interfere with the ordinary traffic ; the rails being laid on iron columns twenty feet high. This railway runs from Thirtieth Street to the Wharves ; and the carriages resemble those of an ordinary railway ; except, that they are lower in the middle than at the ends, thus giving them, it is said, a better grip of the rails. Up to the present no accident has happened, and the system is being gradually extended.

For a city of such size and importance New York is very deficient in fine public buildings. Many of the best edifices appear to have been raised by private enterprise. The finest public building is the Post-office, in the mixed

Doric and Renaissance style, four stories high with a mansard roof, and several louvre domes. It is a noble building, and well adapted to the enormous postal business carried on in it.

The City Hall is a handsome structure in the Italian style, three storeys high, built of marble; the façade being lined with Ionic and Corinthian columns, in the style so prevalent in American public edifices.

The Court House is a plain and massive Corinthian building; but it is not yet completed. It is built of white marble; the main entrance being approached by thirty broad steps, ornamented with massive marble columns. The cost of this building has been $12,000,000; in consequence of the great frauds connected with its erection. " Boss " Tweed is alone supposed to have netted half the total cost.

The Tombs is a granite prison, of pure Egyptian style of architecture, and possesses an imposing but gloomy entrance. In this building the magistrates sit daily, for the trial of minor offences.

The Custom House is a plain building, noteworthy for the immense size of the columns, which support the pediment of the front elevation. Under the dome in the interior is the rotunda; around the sides of which are eight lofty columns of Italian marble with carved Corinthian capitals.

The Treasury is a good specimen of Doric architecture,

built of white marble, and approached by a flight of eighteen marble steps ; it contains a fine rotunda, supported by sixteen Corinthian columns.

The Equitable Life Insurance Office, the Park Bank, the New York "Herald" Office, the "Tribune" Building, the Drexel Building, the Staats Zeitung Building, and the New York Life Insurance Office, are amongst the most prominent, of the private edifices in New York. Stewart's retail store is a fine building, five storeys high, occupying the entire block between Ninth and Tenth Streets, and Broadway and Fourth Avenue.

The Masonic Temple is a fine granite building of five storeys, surmounted by a dome.

New York contains twelve public libraries ; the principal of which are the Astor, the Mercantile, and the Lenox. The first contains 148,000 volumes, and was founded, and endowed, by Jacob Astor, and his son William Astor ; the second, contains 145,000 volumes ; the Lenox is a fine building, but the library has not yet been opened.

The Metropolitan Museum of Art has a fine collection of paintings by the old masters, of statuary, pottery, ceramic-ware, coins, armour, and antique and mediæval curiosities. It also contains the famous Cesnola collection of Cypriote antiquities.

The National Academy of Design is a unique building of gray and white marble, and bluestone ; designed after a palace in Venice, in the Moresque style.   It contains a splendid collection of Egyptian antiquities, and Nineveh relics.   In it is exhibited every spring a collection of the recent works of American artists.

There are in New York 370 churches, belonging to the different denominations; some of which are very fine. Trinity Church, opposite Wall-street, is one of the oldest and most important; it is a fine building, very cathedral-like in appearance.   The land for some distance down Broadway was granted to the Church by Queen Anne in 1705, and has so increased in value, that this Church has become the wealthiest in America.   Attached to it is a churchyard, where repose the remains of Captain Lawrence of the Chesapeake, Emmett the Irish patriot, and other distinguished men.   St. George's Church in the vicinity of Union Square, is a handsome edifice, built of stone, with two lofty towers.   Its interior is very beautiful ; and it is said to accommodate a greater number of people, than any other church in the city.   Grace Church, situated at the sharp turn of Broadway, is the fashionable place of worship.   It is a remarkably pretty structure, and, with its parsonage, forms a nice relief to the monotony of the surrounding masses of brick and mortar.   The Roman Catholic Cathedral of St. Patrick, in course of erection, will be, when completed, one of the finest ecclesiastical edifices in the world, being built entirely of white marble, with beautiful and delicate tracery, resembling in much the celebrated Milan Cathedral.

The Temple Emanuel, the principal Jewish place of worship, is the finest specimen of Saracenic architecture in America. Its internal decorations are magnificent, the colours being beautifully toned, harmonizing well, and a general appearance of richness pervading the whole.

The public institutions of the city, both for educational and charitable purposes, are numerous. The principal educational establishments are the University of New York, a fine marble gothic structure; Columbia College, standing in fine grounds, and containing a library and museum; the College of the City of New York, a free institution, forming part of the national school system; the Normal College, for the training of teachers for the state schools, a fine building in the secular-gothic style; and the Cooper Institute, founded by Peter Cooper. This latter contains a free library and reading room, free schools of art, of wood engraving, of photography, and telegraphy for women; and free night schools for both sexes. It has nearly 3,000 students, and is a most meritorious and praiseworthy institution; in its beneficial effects, perhaps nowhere equalled.

There are numerous hospitals and asylums for the blind, the deaf mutes, and the insane. The principal place for the cure of the sick is Bellevue Hospital, which accommodates 1,200 patients.

A description of New York, however short and limited, would be altogether incomplete without some mention of one of its best features—viz., Central Park. This place of recreation is undoubtedly one of the largest and finest in the world, occupying a rectangular area of 843 acres.

In this park, in addition to several fine lakes, are the two Croton Reservoirs, covering an area of 140 acres; the remaining ground being laid out to form ten miles of carriage drives, and thirty miles of footpaths, and is adorned with numerous bridges, statues, and arbours. The Mall, the principal walk, is bordered by a double row of stately elms ; and contains bronze statues of Shakespeare, and Sir Walter Scott. Central Lake, reached from the terrace, by descending a broad flight of steps, is the loveliest spot in the park ; and the crowd of small yachts and pleasure boats, always to be seen on it, assists in making up a very gay, and animated scene. Central Park is a place of which New Yorkers may justly be proud ; and as public conveyances can always be obtained at a small charge, strangers can easily, and without fatigue, see all its points of attraction.

Some of the theatres are remarkably fine, especially the Grand Opera House, and the Academy of Music. Booth's is handsomely decorated, and Wallack's, and Fifth Avenue are also fine houses. The nine or ten others are second-rate.

The imports of New York amount to seventy-nine, and the exports to seventy-one millions sterling. More than half the foreign trade of the United States, is carried on through this port; and two-thirds of all the Customs duties, levied by the Government, are here collected.

Brooklyn, though a distinct city with separate municipal officers, is generally regarded as a dependency of New York; many of the merchants having their counting-houses in the latter, and residing in the former. It covers an area of twenty square miles, and contains a population of half-a-million. It is often called the "City of Churches" on account of the great number of its religious edifices; and it otherwise possesses many fine public buildings. Prominent amongst the latter, are the City Hall, a marble edifice of the Ionic order; the Court House, another marble building of Corinthian architecture, with a fine portico and dome; the Academy of Design; and the Mercantile Library; the latter, a fine Gothic building, with reading-rooms, and 40,000 volumes.

Prospect Park is a large, and fine recreation-ground, covering 550 acres of ground. The view from here, of the cities of New York, and Brooklyn; of the harbour, with its shipping; and the rivers, with their islands; is very fine. It is beautifully laid out, somewhat after the style of Central Park; but is more densely timbered, and its large meadows, shut in by wooded hills, give it a distinct individuality.

Greenwood Cemetery is however, the lion *par excel-lence* of Brooklyn ; and nothing better illustrates, the decidedly aristocratic tendencies of the better, or wealthier class of Americans ; for here, with a profusion quite unknown in any part of Europe, millions of dollars have been expended, in adorning the graves of the dead.

The Cemetery itself, is a fine park of 413 acres, ornamented with lake, fountains, &c. ; and presenting a varied surface of hill, dale, and plain, traversed by seventeen miles of carriage drives, and fifteen miles of footpaths, shaded by fine trees. The cost of allotments in the Cemetery is so high, that they are only available to the wealthier class ; and as it seems to be the custom, for one family to vie with the other, for the possession of the grandest mausoleum ; and to spend enormous amounts, in trying to attain this end, it may safely be said, that no Cemetery in the world, not even the celebrated Père la Chaise, in Paris, can compare with it, in grand edifices and sculpture. The entrance itself is a work of art, being a monumental structure of brown stone, in the Gothic style, ornamented with sculpture, representing scenes from the Gospels ; the most prominent being, the Entombment, and Resurrection of Christ.

The Atlantic Dock is a fine piece of workmanship. The basin occupies an area of forty-two acres, and the depth of water is sufficient to float the largest vessels. The piers surrounding the basin are of solid granite, upon which are erected large substantial warehouses; and which give a wharfage accommodation two miles in extent. The Dry-dock is another great work ; it is built of solid granite, contains 600,000 gallons of water, and can be emptied by steam pumps in four hours.

A great Suspension Bridge is in course of erection, to connect Brooklyn with New York ; it is to be 6,000 feet in length, the span across the river will be 1,600 feet.

Plymouth Church, of which Henry Ward Beecher is the pastor, is a plain red brick building, but very commodious, seating some 3,000 people. It resembles in appearance a plain theatre ; and under the same roof are rooms, used for lectures, and Sunday School purposes. The Church also possesses other buildings in different parts of the City, and has 2,500 children attending its Sunday Schools. Beecher himself is a hale old man, straight as a dart, and still in evident possession of much of his former energy. His enunciation is clear, and distinct; and the words seem to roll out of his mouth, without apparent effort. He preaches a very practical theology. The Sunday I attended his service, he took as his text, "The night cometh"; and upon this theme he spoke at great length, with special reference to the marriage relations. He treated the text as referring to death, and exhorted his hearers to set their houses in order, to be ready for its dread advent. He referred to the great extravagance in dress, and luxury in living, which compelled the father of the family to leave home early, and return late, in order to earn sufficient to keep up the extravagant style of living, and which prevented him from seeing his children the whole week. He supported with strong arguments the theory that a business which compels a man to labour more than eight hours a-day, should be given up ; insisted upon life insurance, as a provision for the wife and family, in case their bread-winner were suddenly called away ; and suggested other practical matters affecting the relations between man and wife, and father and child. I also visited the Tabernacle, to hear Beecher's great rival Talmage, who is thought by many to be his superior in eloquence; but though much pleased with the sermon, there was not about it that originality that struck me so forcibly in Beecher's.

The heat in New York now became so oppressive, that I went down to Long Branch, for a couple of days, to take advantage of its cooler atmosphere and sea-bathing. Long Branch is a fashionable watering-place, and like all American places devoted entirely to amuse-ment, teems with a mixed, motley assembly of the wealthier class, and of the class, or classes, that prey upon it ; for here congregate a number of parasites, both male and female, swindlers all, who carry on what seems to be a flourishing trade, at the expense of the pleasure seekers, and more especially of those strange to the habits of the country.

Long Branch can lay no claim to rural or marine beauty; the beach is bad, and the town itself consists almost entirely of large temporary hotels and boarding-houses. The mode of bathing is the same as prevails in France. Men and women wear bathing-dresses, and bathe indiscriminately together; and as the beach here shelves abruptly, and the surf is strong, bathers are only able to proceed a couple of yards or so into the water.

Life at Long Branch is circumscribed; visitors rise late, breakfast late, and bathe late; take a siesta during the heat of the day, or attend any races that may be held in the neighbourhood; then a late dinner; and in the evenings there are balls, where they dance late; after which, they sup late; and finally, retire late. I soon got weary of this inane sort of life, and was pleased to return to the heat and dust of New York.

CHAPTER XIII.

BOSTON, HARTFORD, NEW HAVEN, PROVIDENCE.—
NEW ENGLAND CHARACTERISTICS.

NEWPORT — Boston — Harbour — Pride of Bostonians—Public
Buildings — " Common" — Harvard University — Collegiate
Halls — Curriculum—Discipline—City of Hartford—Trinity
College—" Charter Oak" Old Puritan Laws—Position of New
Haven — Yale College — Description of Providence—New
England States—Principal Features—Education—Infanticide—
Irreligion.

TAKING my passage by the steamer *Bristol* of the Fall
River line, I started for Boston.   This steamer, together
with its sister ship the *Providence*, are amongst the largest
and most splendid in American waters ; and there are
few trips more enjoyable than that portion of the journey
to Boston, made in them.

On leaving New York, we had a fine *coup d'œil*,
comprising a grand view of the harbour ; of the cities of
New York and Brooklyn ; of the shores of Long Island,
and the numerous small islands in the East River.   The
passage was nearly all the way through the calm waters
of Long Island Sound, and the Atlantic was only gained,
just before the steamer entered the Fall River, whence the

railway cars soon conveyed us to Boston. We made one stoppage before arriving at the Fall River at Newport, a clean little town in Narrangansett Bay, and a place of great resort in the summer months. Here is an old ruin, called the Northmen's Tower, supposed to have been built by the early Norse discoverers of America. This may or may not be, but the tower with its rough piers and capitals has a very ancient appearance, and is, in any case, a very curious relic.

Boston, the capital of the Old Bay State as Massachusetts is sometimes called, and chief city of New England, is situated on Massachusetts Bay and comprises Boston proper, East Boston, South Boston, Roxbury, and Dorchester. The city proper covers a peninsula only 700 acres in extent, and is consequently very closely built upon.

East Boston occupies Maverick's Island, and having the greatest depth of water, at its wharves lie all the ocean steamers, and vessels of large draught. Some of these wharves are of considerable extent, that belonging to the Cunard Company being over 1,000 feet long.

South Boston extends for about two miles along the south side of the harbour, by an arm of which, it is separated from Boston proper.

The city is connected with Charlestown, one of its suburbs, by the Charles river bridge 1,500 feet long ; and with Cambridge, another of its suburbs, by the West Boston bridge 2,750 feet in length.

The harbour covers an area of about seventy-five square miles, and is studded with numerous islands, which give it a very picturesque appearance. It is memorable for having been the scene of the occurrence, that led to the battle on the neighbouring heights of Bunker Hill, the actual beginning of the Revolutionary War, the results of which have been perhaps more important, than those of any other historical event. This occurrence, as is well-known, was the destruction, by a party of Bostonians disguised as Indians, of the obnoxious tea, attempted by the Home Government of the day to be forced upon the colonists.

Boston is one of the oldest cities in the States, having been founded in 1625, a few years after the settlement of New York. In 1872, it was the scene of an immense conflagration, which caused great devastation, destroying about 800 of the best buildings, which have however in an incredibly short space of time, been replaced by other, and better edifices. It is very much like an English provincial city in appearance, and is without doubt the most old-fashioned town in the States; the streets being mostly narrow and crooked, especially in the older part.

Washington and Tremont Streets are the principal business thoroughfares, and contain many fine shops, and other buildings. Since the fire the streets generally have been widened, and otherwise improved.

Boston has often been called the " Athens of America," from Harvard University, the most ancient seat of learning in the country, being in its immediate vicinity ; and from its possessing many other institutions devoted to higher learning. The pride of its New England inhabitants is proverbial, and was exemplified after the great conflagration ; when Chicago and other towns collected by voluntary contribution, and forwarded a sum of £40,000, for the use of the people rendered homeless ; which the Bostonians refused to accept, on the plea that they were rich enough to support their own poor.

The public buildings of Boston are neither remarkable for their number, nor beauty ; the most prominent are the State House, and the City Hall. The former is a building in the Grecian style of architecture, with an imposing colonnade in front, and surmounted by a gilded dome. The entrance leads directly into the Rotunda, in which are a fine statue of Washington, busts of former governors, and trophies of banners and cannon. The City Hall is an imposing building of white granite, in the Italian Renaissance style, with a fine louvre dome. On the lawn in front is a fine bronze statue of Franklin.

The Custom House is a stately granite structure, built

in the form of a Greek cross ; the portico on either front
being supported by heavy Doric columns. The new
Post Office in course of erection will be the finest
building in New England. The Masonic Temple contains handsome Egyptian, Corinthian, and Gothic
Halls.

The Boston Public Library is a very meritorious Institution, exceeded only in size and importance, by the
Library of Congress at Washington. It contains 260,000
volumes, 100,000 pamphlets, and the celebrated Tosti
collection of engravings. The library and reading-rooms
are open gratuitously to all, and residents have the
privilege of taking books home.

The Athenæum is one of the best endowed in the
New World, and contains galleries of sculpture and
paintings, and a library of 100,000 volumes.

The Museum of Fine Arts is a new building of red
brick, elaborately ornamented with terra-cotta bas-reliefs,
and contains some of the most valuable works of art in
the country.

The Institute of Technology is a school for instruction
in the applied sciences, and for granting degrees in
engineering ; it is on the model of the German technical
schools, and is found to be most beneficial in its results.
Similar schools are now being introduced into the
Australian Colonies, and will, when properly established,
doubtless prove of great benefit.

Boston contains several old buildings famous for memorable events connected with the Revolutionary times ; the principal of these being Faneuil Hall, called the " Cradle of Liberty."

Right in the heart of the City is the " Common," an area of 48 acres, laid out in lawns and walks, shaded by fine trees, and surrounded by a handsome iron railing. This park, in so densely populated a city, is a great boon to the people, and an incalculable benefit to the public health. Here in the evening numbers of the citizens are to be seen, enjoying the fresh air ; and the " Common" then presents quite a gay appearance. Portion is used as a public garden, and is beautifully arranged, and adorned with several fine statues.

Mount Auburn Cemetery, after Greenwood, the finest in America, occupies 125 acres of ground ; is well laid out, and contains numerous artificial lakes, costly monuments, and fountains. The entrance, built in the Egyptian style is remarkably fine.

At Charlestown, is the Bunker Hill Monument ; a plain massive obelisk 221 feet high, commemorating the great battle fought on June 17th, 1775.

Harvard University, the most important seat of learning in the States, is at Cambridge, a suburb of Boston ; it was founded by the Rev. John Harvard an alumnus of Emmanuel College, Cambridge, England, and now comprises, besides the collegiate branch, schools of divinity, law, medicine, dentistry, and the Lawrence school of science. The college-yard, 15 acres in extent, is a fine green with numbers of stately elms ; and within it are clustered the fifteen plain and unpretentious buildings, forming the University. One of these buildings, however, the Memorial Hall, used as the Senate House, is a beautiful stone structure, with a fine tower 200 feet high. It was raised at a cost of £100,000 by the alumni and friends of the University, in honour of those students who fell in the Civil War ; and contains a large theatre or lecture-hall, and various other large halls, the principal of which, has splendid stained-glass windows, and on its walls are carved the names of the alumni in whose honour the edifice was erected. Other college buildings are Matthews, Gray and Boylston Halls, used as dormitories ; Massachusett Hall, as a reading-room ; Thayer Hall ; Dane Hall, used as the law school ; and Gore Hall, which contains the University library of 130,000 volumes ; besides which there are an Observatory, a Zoological Museum, Herbarium, and Gymnasium.

The undergraduates number nearly a thousand, the majority of whom are accommodated in the various dormitories, though many sleep in the town.

The curriculum resembles that of Oxford, prominence being given to classical studies, which causes Harvard to have a distinct individuality amongst American higher educational institutions ; the tendency generally being towards utilitarianism. The matriculation examination, which candidates for admission have to pass, is very difficult, and the whole course of study is stringent : a student having to attain a high state of efficiency before he can obtain a degree. The discipline, however, is very lax, there being no restrictions outside the class-rooms ; undergraduates being allowed to do pretty much as they like. The students are drawn from the better class, and seem to be a gentlemanly lot of men ; they come chiefly from the New England States. Southern and Western Americans generally prefer the great kindred institution of Yale College, on account of its more orthodox Christianity.

It struck me, that at Harvard, the mind is trained at the expense of the body ; as the undergraduates do not appear to go in for much exercise, and with the exception of a little boating, and the national game of base-ball, there do not seem to be any outdoor sports. I would say, however, that a graduate of Harvard should be a thoroughly well educated man ; and the institution, in spite of a few drawbacks, is one of which Americans may justly be proud.

After spending a few days in viewing the lions of Boston, I left by the New York and New Haven Railway for Hartford, which, until last year alternated with New Haven as the capital of Connecticut, but which has now been made the sole metropolis. It is one of the

prettiest of the many pretty New England towns, lying
in the valley of the Connecticut River, at the head of its
navigation and junction with Park River ; the country
in the vicinity being rich, cultivated plains, interspersed
with well-wooded hills, and dotted with villages and
homesteads.

The town itself is regularly laid out, some of the
streets running parallel with the Connecticut river and
being crossed by others at right angles ; it is divided
into two portions by Park river, which is spanned by
numerous bridges.   Main Street, the principal thoroughfare, presents for about a mile an unbroken front of
shops, public buildings and churches.

The city proper is connected with its suburb East
Hartford, by a fine iron bridge over the Connecticut
river 1,000 feet long.   Hartford boasts a fine public
park, covering forty-six acres of ground, prettily situated
on the banks of Park River ; at the upper end of
which, are the buildings of the Episcopal College of the
Holy Trinity, the leading educational institution of the
city, and one of the most important in the State.   The
course of instruction at Trinity College is similar to that
at Yale ; but more attention is paid to religious studies.
There are about a hundred students in residence, and a
staff of about twenty able professors.   Divine Service is
held twice daily, at which the attendance of the undergraduates is compulsory ; and attached to the College is a

fine Library of 7,000 volumes, where the mitre, which belonged to Bishop Seabury, is preserved as a relic,— such ornaments not being used in the American Episcopal Church. The College buildings are three in number, and are plain, substantial stone structures ; but a new site for the College has been purchased, and it is intended to erect better and more suitable edifices.

In close proximity to Trinity College, is the new State House, in course of erection : this is to be a large marble Gothic structure, and will contain in addition to the two Chambers of the Legislature, the State Library, and the Supreme Court.

The State Library contains the original Charter, granted by Charles II. to Connecticut, which was for some time concealed in an oak tree, from the wood of which a chair has been made, which now stands in the Senate Chamber. I was shown copies of some of the old laws enacted by the Puritans, which assuredly do not err on the side of leniency ; but prove these old zealots to have been narrow-minded and bigoted to a degree.

Disobedient sons were punished with death, and people convicted of being Quakers had one ear cut off for the first conviction, the other for a second ; whilst for a third, they were to have their tongues bored with a red-hot iron, and to be whipped.

The Asylum for the Deaf and Dumb, situated in the midst of extensive grounds, is a most meritorious, and

well-conducted institution, and was the first of the kind
erected in America.

Another noteworthy institution is the Wadsworth
Athenæum, a fine castellated edifice, containing a gallery
of paintings, and statuary, a museum, and no less than
three distinct libraries, belonging to different associations.

The population of Hartford is about 40,000 ; and it is
the centre of numerous, and important manufactures, the
principal being the fabrication of iron and brass-ware,
steam-engines and boilers, firearms, tools, sewing-machines,
and plated-ware.

Just outside the City, on the river bank, and forming
a village by itself, is Colt's large firearm factory ; and in
close proximity, is the pretty little Episcopal Church,
erected by Mrs. Colt for the use of the men employed in
the works.

A short journey of thirty-six miles, through a good
agricultural country, past several clean little towns, brought
me to New Haven, the largest and most populous city of
Connecticut, and one of the oldest settlements in New
England. This city is built on a plain sloping gently
down to New Haven Bay, with a background of wooded
hills, two rocky promontories called East and West Rocks,
especially forming a striking feature in a picturesque
landscape. The town itself is clean and cheerful looking,
and contains a population estimated at 55,000 souls.

Chapel Street, containing the principal business and

public buildings, extends right through the city; whilst the private residences are in avenues, lined with stately elm trees; for which reason New Haven is often called the "City of Elms."

In the middle of the town is the "Green," a square shaded by fine trees, fronting on which are several churches, and the City Hall, a handsome building, containing in addition to the municipal offices, a large hall in which courts are held.

New Haven is the centre of a large railway system, and possesses numerous industries; the manufacture of machinery, clocks, firearms, hardware, and pianos, being the principal.

Near the "Green" is College Square, in which are the old fashioned buildings comprising the celebrated Yale College, the great rival of Harvard University. The Library, the Theological School, and the modern Durfee and Farnum Halls, are certainly improvements upon the other ugly buildings, which have, however, an old-fashioned appearance, and air of comfort, very rare in America. Connected with the College, are schools of divinity, law, medicine, science, and the fine arts. The Library contains 90,000 volumes; and the Art building has some fine collections.of paintings and casts from Greek antiques. The Old Commons' Hall contains fine geological collections. The School of Science has a laboratory, library, and scientific collections.

Alumni Hall is adorned with numerous portraits of distinguished graduates. Yale was founded in 1700, and is consequently nearly as old as Harvard ; it has now over a thousand students in residence. The course of instruction extends over four years, and is very severe ; whilst the discipline outside the class-rooms, as at Harvard, is very lax. Every student is compelled to attend morning service, which is however a very cold affair ; the prayers being mumbled over, whilst the students, all the while, are preparing for the lectures of the day.

Leaving New Haven by the Stonington and Providence Railway, the road lies for some distance along the shores of Long Island Sound, through well-cultivated country; and after passing the important towns of New London, Stonington, and Westerly, we arrived at Providence, one of the capitals of Rhode Island, the smallest of all the States in the Union. This city is situated on an arm of Narrangansett Bay, called Providence River, and is, after Boston, the most populous and wealthy city in New England. It is a remarkably pretty place, the prettiest in fact of all the New England towns. The river flows into the centre of the city, where it expands into a fine lagoon, nearly a mile .in circumference ; and round this a park has been formed, planted with beautiful elms ; in addition to which there are numerous greens or squares, and a pretty park of 100 acres, called after the founder of the city Roger Williams, by a descendant of whom it was presented to the citizens.

The land upon which Providence is built, is somewhat hilly; and therefore unlike the flat monotony of New Haven, it presents a diversity of appearance, rendered picturesque by the hills being covered with fine residences. The principal business thoroughfare is Westminster-street, and from it extends the Arcade, one of the lions of the city, of which the people of Providence are very proud. This Arcade is 225 feet long and 80 feet wide, and has three stories containing shops; it is entered at either end through an imposing Doric portico. On the heights at one end of the town, are the six buildings of the Brown University, an old, and one of the best educational institutions in New England. These buildings stand in extensive grounds, and contain a large library and art collection.

The public buildings of Providence are neither remarkable for their number, nor beauty ; several now in course of erection, when completed, will materially add to the architecture of the city. In the centre of one of the squares, stands a very fine monument of granite, with five large bronze statues, erected to the memory of those Rhode Island soldiers and sailors, who fell in the Civil War. On this monument are inscribed the seventeen hundred and odd names of those it commemorates.

There are seventy-six churches in the city, some of which are fine buildings.

The Athenæum is a very useful institution, containing a good library and reading-room, and a collection of paintings. There are also several meritorious establishments for the cure of the sick, and the relief of the distressed.

Providence contains a population of over 100,000, and is the most important railway centre in New England, and the seat of many manufactures; the principal being cotton and woollen goods (it being the leading American market for "prints"), the Gorham-plate, the Peabody rifles, and the celebrated Corliss engines.

New England comprises the six States of Maine, New Hampshire, Vermont, Connecticut, Rhode Island, and Massachusetts; and contains a gross population of 3,500,000 inhabitants. These States were originally settled by the Puritans, who sought in the New World that liberty of conscience, and freedom of thought, denied them in the Old. How far, and to what extent the principle of toleration was carried out by them, may be judged by a perusal of some of their laws and enactments given in an earlier part of this chapter.

The country has been brought into a high state of cultivation, and the towns, though possessing a certain primness, are clean, and pretty, and have a more settled appearance than the majority of American towns; their environs especially, being tidy and well kept, and having in place of the wooden shanties so common in Western towns, nice villas and trim gardens.

Great attention has been paid to the education of the people and the chief seats of learning have their home here; the taste too for the Arts and Sciences is more cultivated, than is the case in other States of the Great Republic.

New England has hitherto given the tone to American society and general national characteristics; but this privilege is being gradually lost in the rising greatness of the Middle, Western, and South-Western States.

Life here, where the natural obstacles have long been overcome, is quieter, and more peaceful than in the West, where man is still waging a war with nature to bring the soil under cultivation; and there is a consequently smaller number of crimes of open violence, although infanticide is said to prevail to a frightful extent. Researches prove the fact, that it is owing to the prevalence of this crime, that the descendants of the original Puritan settlers are fast decreasing in number.

Irreligion is general; Unitarianism prevails to a great extent; whilst Universalism counts many adherents; and it must not be forgotten that New England has given birth to many of those free-love doctrines, that still like fungi, cumber the earth. Such is the outcome of Puritanism.

CHAPTER XIV.

PHILADELPHIA AND THE CENTENNIAL EXHIBITION.

CITY of Philadelphia—Fairmount Park—Squares and Streets—
Independence Hall—Other Prominent Edifices—Girard College
— Exhibition Buildings — Classification of Exhibits — Main
Building—Machinery Hall—Agricultural Hall—Memorial Hall
— Horticultural Hall — Women's Pavilion — Government
Building — Subsidiary Buildings — Exhibition Grounds — Centennial Fountain—Accommodation for Visitors.

RETURNING to New York, I proceeded by the Pennsylvania Railway to Philadelphia ; the journey being through a flat and uninteresting country, but past many important towns, amongst others, Newark a city of 100,000 inhabitants.

Philadelphia, the old Quaker City of Brotherly Love, as its name implies, is the largest in the States, in point of area, and the second in population. It is situated between the rivers Delaware and Schuylkill, about six miles above their junction ; at present contains about 350 miles of paved streets, and a greater number of houses than any other city in America. It is very regularly laid out, the streets all running north and south, or east and west ; the former are numbered successively from the Delaware to the Schuylkill, commencing with

First Street, and going up to Twenty-third, and as the city has outgrown its former limits, they are now continued on the other side of the Schuylkill and proceed for a great distance. The houses in the streets running east and west, which have mostly pomological names, as Chestnut Street, Walnut Street, Vine Street, Filbert Street, &c., are all numbered from east to west; all between First and Second Streets, being numbered from 100 to 200, and all between Second and Third Streets, from 200 to 300, and so on. This is a system that might advantageously be carried out in all new cities, as it enables a stranger to find his way about very easily. For instance, if the nearest house be 940 then the visitor knows he is between Ninth and Tenth Streets; and in like manner, as all the numbered streets running north and south are allowed a hundred numbers, for each block distant from Market Street, the centre of the city; a stranger has only to learn the cardinal points, and he can calculate to a nicety, his distance from the Delaware or Market Street. The streets however are not wide, and in spite of their regularity, Philadelphia cannot be called a fine city; especially as its public buildings are not so imposing as those of many less important places. Like Boston, it much resembles an English city, and the inhabitants generally have an old-country look.

The great feature of Philadelphia is Fairmount Park, one of the largest in the world, covering a space of

2,740 acres, extending for seven miles on both sides of the Schuylkill River, and Wissahickon Creek. It is nearly fourteen miles in length, and possesses great natural beauty; though up to the present time, little has been done towards its improvement. Within the park are situated the reservoirs that supply the city with water, which latter is brought from the Schuylkill, and forced up to the reservoirs by hydraulic power. Fairmount also contains a colossal bronze statue of Abraham Lincoln, and a small Art gallery in which are a few very fine pictures.

Adjoining, is the new Zoological garden, which has only recently been formed, but which already contains a good collection of animals. The grounds are well laid out, and the various houses for the animals are very elegant structures.

There are some pretty squares in the city. Logan Square contains seven acres, nicely arranged; Washington Square is enclosed by a handsome iron railing, is very well kept, and contains a specimen of every kind of tree that will grow in this climate, whether indigenous or not. Franklin Square has a fine fountain in the centre, and has a very trim appearance. Rittenhouse Square is in the aristocratic portion of the city, and is surrounded by fine mansions.

The principal streets are Chestnut Street, which contains the best retail establishments; Market Street, the principal business thoroughfare; and Broad Street, containing the principal private residences and churches. Other leading thoroughfares are Lombard, Arch, Race, Vine, and Third Streets.

The most interesting building in Philadelphia is undoubtedly Independence Hall; in it the first Continental Congress was held ; and here, on July 4th, 1776, the Declaration of Independence was adopted, and publicly proclaimed from the steps. The room in which Congress sat on that memorable occasion, presents the same appearance now, that it did at that time ; the furniture having all been preserved. The building also contains a statue of Washington, numerous portraits of celebrated personages connected with the passing of the Declaration of Independence, and many curious revolutionary relics. Here also is kept the celebrated Liberty Bell, the first one in the United Colonies that rang out a peal of joy, after the proclamation of Independence.

The public buildings are at present unpretentious, but there are now in course of erection a new Post-office, to be in the French Renaissance style, and on a very extensive scale, and a new building to be occupied as Law Courts and Public Offices. This edifice is to be of white marble, 486 feet long by 470 wide, four stories high, and will cover an area of four acres and a half, exclusive of a court-yard 200 feet square; the centre tower will be 450 feet high.

The present Post-office is a plain white marble building. The Custom-house is a good specimen of Doric architecture, modelled on the Parthenon at Athens, with the exception of the columns at the sides. The Mint is also a fine white marble building in the Ionic style.

The Eastern Penitentiary covers about ten acres of ground, and in appearance resembles a baronial castle of the Middle Ages; here the silent system is in force, prisoners being allowed to speak with the chaplain and prison officials, but not with their fellow-prisoners. Sufficient work is provided to keep them occupied.

The Merchants' Exchange is a large marble edifice, with a semi-circular colonnade of eight pillars, and a spacious Rotunda, which contains a handsomely-frescoed reading-room.

The United States Naval Asylum is a large marble building, standing in the midst of extensive grounds, and has a fine Ionic portico with eight graceful columns. The city also possesses two Arsenals, one of which is devoted to the manufacture of army clothing, shoes, &c. ; and the other to the storage of ammunition.

The Masonic Temple, the finest in the world, is a solid granite structure, in the pure Norman style ; it is richly decorated in the interior, and contains large Corinthian, Doric, Egyptian, Ionic, Oriental, Norman, and Gothic Halls.

Foremost amongst the Educational establishments is

the University of Pennsylvania, which is one of the oldest institutions in the country, having been founded in 1749. It occupies a group of handsome buildings, and contains a fine library and museum, and gives instruction to eight hundred students.

Girard College is a grand institution, founded by Stephen Girard, who left two millions of dollars, to be devoted to the erection of suitable buildings, to provide for "the gratuitous instruction and support of destitute orphans"; and the residue of his large estate he devised for its maintenance. All honour to this great and generous man! The College grounds extend over an area of 42 acres; the College, erected on a height, is a noble marble building of the Corinthian order of architecture, in the form of a Grecian temple; it is surrounded by thirty-six marble columns, and is one of the most chaste and beautiful structures in America. It at present contains 540 inmates, and from its roof, built of marble, and rising in steps from eaves to ridge, a most commanding view of the city is obtained. On the grounds is erected a fine monument to the graduates of the College, who fell in the Civil War. In his will Girard provided that the orphans should be fed with wholesome food, clothed with plain, decent apparel, and instructed in the branches of a good sound education. It is peculiar that while secular visitors are allowed to inspect the College, no clergyman of any

·denomination is admitted; it having been especially provided in Girard's will, that he desired to keep the ·tender minds of the orphans free from the excitement, ·which sectarian controversy always engenders.

Amongst other high-class Educational institutes are the Hahnemann Medical College, the Wagner Free Institute -of Science, the Polytechnic College, and the College of Physicians.

Philadelphia contains numerous charitable institutions; ·the Pennsylvania Hospital, and the Hospital for the Insane, are both admirable in their way.

Of the 420 churches in the city, many are fine build-ings, and there are numerous libraries, scientific institutes, .and other associations and societies, devoted to various ·purposes.

Philadelphia has little left of its Quakerism, except it ·be the formal manner in which it is laid out; it has, on ·the contrary, the unenviable reputation of being one of the fastest of the Eastern-American cities, and is notorious ·for dissipation and crime.

It may be remarked as a curious coincidence, that on ·the breaking out of the war between Great Britain and ·the United States in 1812; a whale ascended the Dela-ware as far as the city, where it was caught. This ·was a hitherto unprecedented circumstance, which was however, repeated in 1861, on the breaking out of the ·Civil War.

At the time of my visit to Philadelphia, the great Centennial Exhibition, commemorating the Centenary of American Independence was being held, and a few words about it, may not be out of place. The reason why Philadelphia was selected as the place for holding the Exhibition was, that the Declaration of Independence was there signed and proclaimed to the people, and there also the first Congress met. Boston and New York both put in a claim for the honour, but it was rightly accorded to Philadelphia.

The Exhibition was held in a portion of Fairmount Park, allotted for the purpose, containing about 230 acres. The articles for exhibition were classified in seven departments, and placed in five distinct buildings as under :—

1. Mining and Metallurgy ⎞
2. Manufactures .. ... ⎬ MAIN BUILDING
3. Science and Education ⎠   Covering an area of 21½ acres.
4. Art ... ... MEMORIAL HALL ,, 1½ ,,
5. Machinery ... ... MACHINERY HALL ,, 14 ,,
6. Agriculture ... ... AGRICULTURAL HALL ,, 10¼ ,,
7. Horticulture ... ... HORTICULTURAL HALL ,, 1½ ,,

              —————
              48¾ acres.

In addition to the above main buildings there were also :—

GOVERNMENT BUILDING, for " the purpose of exhibiting such articles as tend to show the functions and administrative faculties of the U. S. Government in time of peace, and its resources as a war power."

THE WOMEN'S PAVILION, erected by subscriptions from the women of America, for the purpose of showing articles of female work only.

BREWERS' HALL.

PHOTOGRAPHIC HALL.

CARRIAGE ANNEXE, and numerous other smaller buildings, mostly private, which, with the State Houses and Foreign Commissioners' Houses, covered together an area of 26 acres.

The buildings used for Exhibition purposes, thus altogether covered an area of 75 acres, being 25 acres in excess of the last Vienna World's Fair.   Unlike that Exposition too, where the principal buildings had been circular in form, the Americans at this, their great Centennial Show had constructed theirs rectangular in shape.

MAIN BUILDING.

This structure constructed principally of glass and iron, was in the form of a long parallelogram ; its great length and unbroken front, which might otherwise have given it the appearance of monotonous sameness, being relieved by numberless small towers.   The larger portion was one story high; the interior height being 70 feet ; whilst towers 75 feet high were erected at each corner.; In its construction the building combined lightness and stability in a marvellous degree, and was in itself an exhibition of the skill of its architect and builders.   The roof, which was in three spans was remarkable for strength and simplicity, and tended to prove that mere weight in iron-work is not a gain but a positive disadvantage. The length of this mammoth building was 1,880 feet; its breadth 464 feet; and in its construction $1,600,000 dollars had been expended. The manner in which the floor-space was divided was as follows. Traversing the entire length from east to west were five avenues, each over a third of a mile in length, and the centre one being 120 feet wide. These were intersected by broad thoroughfares, all crossing one another at right angles, and forming a total length of five miles; which it was necessary to traverse in order properly to inspect the exhibits. In the centre of the hall a large space had been left and furnished with galleries, from which a fine bird's-eye view of the whole contents of the building could be obtained. The panellings over this open space were adorned with emblematical designs, representing the four great divisions of the globe. On one side was represented Europe, pouring its treasures at the feet of Shakespeare and Charlemagne. On another square was depicted Asia, its representative men being Mahomet and Confucius. Africa was represented by Rameses and Sesostris, with some very inscrutable looking sphinxes, and America by Washington and Franklin.

It is quite impossible, in a limited space, to describe the varied and splendid display in the main building,

13

which constituted the principal portion of the Exhibition. In it were collected the choicest treasures of science, and the greatest triumphs of manufacturing art; illustrating what a high degree of perfection human industry and ingenuity have attained.

All the principal Courts abutted on the main avenue, and extended back to the wall; broad passages forming the divisions between the different countries.

Commencing at the right-hand side of the main avenue, immediately upon entering, the Italian Court first claimed attention; not by reason of its position alone, but also on account of the great attractiveness of its display. Here were to be seen exhibits of those articles in which Italy has obtained well merited pre-eminence; beautiful mosaics and cameos, delicate Venetian glassware, elegant jewellery, filagree work, bronzes, carved furniture, and church ornaments. Intermixed with these *articles de luxe* might also be seen many things of more homely manufacture, such as enter more into every-day consumption. This Court proved a great source of attraction to visitors, and always seemed to be crowded with sight-seers.

Next in order was the Norwegian Court, which contained a good display of thick woollen and cotton cloths, of carved furniture, and wood work, of glassware, of silver plate, of soaps and perfumes, with trophies of cod-liver oil bottles, etc.

In close proximity was the Swedish Court, containing a splendid collection of rich furs, some beautiful specimens of filagree work, fine wood carvings, a large assortment of arms of all descriptions, numerous articles of Bessemer steel manufacture, and some excellent samples of cutlery. A striking feature in this Court, were groups of life-size figures of the Swedish peasantry in their national dress. A family group contemplating the death of a stag, that had just been shot by the two males of the party, especially attracted attention.

Adjoining the Swedish were the Australian Courts, which, though not so attractive as the majority of the other sections, yet had a certain individuality about them, that quietly yet forcibly claimed attention. The exhibits collectively well represented the great, and as yet hardly developed resources of the country, and in their way were unrivalled. The fineness of the samples of wool shown, the quality of the grain, the large yield of gold, the excellence and quantity of the copper and coal, the native woods, and the fine grain of the sugar, especially taking the Americans by surprise; and it must be satisfactory to Australians to know, that discussions have already 'arisen in the American papers, as to the desirability of modifying their tariff in such manner, as to remove the restrictions on the importation of Australian wool, which they acknowledge to be finer than any that can be produced in the States, and very suitable to their

woollen manufactures. The exhibits of the different
Australian Colonies individually may have appeared
somewhat insignificant, as compared with other sections ;
but the whole display taken collectively, was one that
seemed very much to astonish the majority of the
visitors, whose ideas of Australia seemed hitherto to
have been of a very vague nature. This exhibition
must eventually be of inestimable value to the whole
group of colonies ; for it has materially spread the
knowledge of the great resources of this, the least
known quarter of the globe. As might be expected, the
most admirable portion of the Australian display, and
the one that most strikingly showed its great natural
sources of wealth, were those exhibits representing the
great industries, such as the wool, grain, wines, and
mineral samples. The articles of colonial manufacture
did not compare at all favourably with the exhibits of a
similar kind of older countries. In the Victorian Court
especially, many of the manufactured articles exhibited,
seemed to have place there for the sole purpose of filling
up surplus space. It was remarked too, by many, that
the exhibits of Victorian manufacture, in spite of the
so-called advantages of protection, were not equal in
finish to, or so cheap in price as, those of free-trade
New South Wales. This fact must speak volumes in
favour of unrestricted trade.

The samples of saddles and harness exhibited in the

New South Wales Court received much attention, and it has been remarked by persons connected with that trade in America, that it would be useless to think of exporting to Australia, if such splendid articles in that line can be made there. The mineral wealth of the Australian Colonies was well represented by great trophies representing the quantity of gold exported, by rich specimens of quartz, heaps of smelted tin, pyramids of coal and kerosene shale, and by masses of iron ore and copper. The products and manufactures exhibited comprised numerous samples of wool, grain, silk, sugar, arrowroot, preserved fruit, confectionery, wines, spirits, liqueurs, and vinegar ; flour, biscuits, preserved meats, tweeds, and blankets ; silver goods, furs, rugs, and dressed sheep-skins ; stuffed birds and animals ; samples of olive and eucalyptus oils ; brass and iron castings ; pianos and billiard tables ; samples of leather, and numbers of other articles. The numerous and well executed photographs of the scenery, and cities of the different colonies, must have tended materially to give visitors to the section a striking idea of Australian advancement.

The various courts of the remaining British Possessions came next. The Indian exhibits were few in number, and the display generally fell far short of what might have been expected from that great appanage of the British Crown. The articles exhibited however were very beautiful, and consisted of rich embroideries and shawls ; silks, sandalwood boxes, ivory and feather fans ; lacquered work, ivory carvings, stone and metal work, and other Oriental *articles de luxe.*

The Cape of Good Hope only exhibited samples of its principal industries, prominent amongst which were to be seen a huge pyramid of wine bottles, specimens of wool, minerals, ivory, and skins, and many others representing its various products. The wall space of this Court, like the Australian section, was covered with photographs, representing South African scenery and towns.

Jamaica had a modest display of its woods, rum, and sugar ; whilst Bermuda, the Bahamas, and the other smaller British Dependencies were represented by unassuming collections, consisting of the various products of their different latitudes.

Canada had evidently used every endeavour to make a display worthy of itself, and certainly succeeded in doing so, for the number and excellence of the articles exhibited in this section, excited the admiration of all visitors to it. In manufactures, the exhibits comprised tweeds of fine finish, leatherwork, iron and stone work, implements and tools, articles connected with ship-building, furniture, and furs, etc. ; all displaying great taste and finish, and showing how far Canada has advanced in the mechanical arts, and what a high state of development has been reached in many departments of manufacture. The mineral wealth of the Dominion was also represented, and its mineralogical collection was so well arranged, that it claimed universal attention, Here were to be seen columns of beautifully grained granite, huge blocks of plumbago, pyramids of coal, iron ores, with samples of the manufactured articles; slabs of marble, gold in quartz and slate, and a fine display of mineral oils. Altogether the exhibits of Canada were numerous, and well arranged, and need not have feared comparison in any way, with those of its great neighbour, and rival, the United States.

Leaving the Canadian Court, the section of Great Britain and Ireland was entered, and here a display, though acknowledged to be not so good as it might have been, was yet of so bewildering and dazzling a nature, on account of the multiplicity and excellence of the articles exhibited, as fully to justify Great Britain's claim to industrial pre-eminence. It would require a volume to enumerate and descant upon the variety and beauty of the exhibits in this Court, which comprised every department of manufacture, and exemplified in a high degree, how the genius of man has made the products of nature subservient to his will, and causes them to minister to his wants. I heard it frequently remarked, that the British Court claimed attention and admiration, on the score alone, of the excellence and high finish of the exhibits; there having been, for so large a collection, a singular absence of many articles attractive in themselves, and which would have served to enliven the display of the more useful articles. Jewellery for instance was conspicuous by its absence, and Elkingtons were the only exhibitors of plated ware; although their display, which fronted the transept or main avenue, was in itself sufficient to prove Britain's superiority in that branch of manufacture. Pins, needles, reels of cotton, tweeds, prints, iron and steel work, stone and earthenware, tools, hardware, paper and the thousand and one useful articles exhibited by firms of world-wide reputation, undoubtedly represent great industries and enter largely into the consumption of every day life; still they do not as a rule prove attractive to general sight-seers, and on that account the French and even the Italian Court, always seemed to be more crowded with visitors, than the great British section. The display of beautiful laces, and needlework, of rich silks and splendid upholstery, of magnificent furniture, of elaborate carpets and elegant majolica-ware, scarcely seemed to relieve the general monotony of the whole. The flat printing press, exhibited by the proprietors of the *Graphic*, which was in daily operation, was a great centre of attraction.

Next in succession to the British, came the French Court, which was very much smaller in size, but in which the exhibits were arranged in a very attractive manner, and consisting as they did, for the most part, of the numerous articles which go by the generic term of "*articles de Paris*," were much admired by visitors and the Court always seemed very crowded. This is to be attributed to the more attractive nature of the goods displayed, prominent amongst which were to be seen elegant jewellery, beautiful bronzes, masterpieces of the milliners' art in the form of ladies' dresses, silks, velvets, majolica-ware, gloves, fans and all those *articles de luxe*, for the production of which, the French enjoy such world-wide fame. Many of the French exhibits bore a striking resemblance to those of a similar description in the British Court, and this was particularly noticeable in the specimens of majolica-ware, plated-ware, silks and velvets, porcelain goods, cutlery and watches.

Beside the great and gorgeous collection in the French Court, and almost eclipsed by it, came the modest Swiss exhibits, which however, in point of excellence of finish and lowness of price, could challenge comparison with those of any other section. They consisted of watches, musical boxes, mathematical and surgical instruments, clock materials, tools, lace curtains, splendid wood carvings, textile fabrics, straw plaitings, and school apparatus, the latter showing what a high standard of education prevails in the little Republic.

Belgium came next in order, and its exhibits evidenced the substantial prosperity that exists in that country, and the high position it occupies amongst industrial countries, in many departments of manufacture.

The principal articles exhibited were samples of iron and steel manufactures ; but there was, in addition, a great display of the celebrated Brussels and Mechlin laces, of woollens and linens, carpets, silk embroideries, marble mantel-pieces, and numerous other exhibits, all of high finish and excellent design. A beautifully carved pulpit and canopy had many admirers, and was the *piece de résistance* in this court.

Brazil, which next claimed attention, seemed to have made a great effort to show a worthy display at this World's Fair ; and had erected a most gorgeous pavilion, resplendent in all the colours of the rainbow, in which to display its various exhibits. These represented the products of the empire over which Dom Pedro reigns, and consisted principally of numerous samples of coffee, rice, sarsaparilla, cocoa, ginger, various barks, woods, and specimens of gold and ivory, both manufactured and otherwise. The remaining space in the court was filled out with stuffed birds, artificial flowers, and many articles of native Indian manufacture.

The small Court of the Netherlands, which was in close proximity to that of its neighbour, Belgium, was crowded with a heterogeneous collection of articles, both of use and luxury ; and many hours were requisite, properly to note and admire its many excellent exhibits. These were of the most varied description, and con-sisted in part of rich Delft carpets, woollens of beautifully soft texture, silk fabrics, jute manufactures; tin, iron, and wooden ware; elegant screens, papier-maché articles, samples of woods and metals; and numberless articles, representing various other industries. A conspicuous feature in this section was the collection exhibited by the Artizans' School of Rotterdam, an institution founded in 1869 for lads of twelve to fifteen to be "practically and theoretically trained to become clever artizans."

The Mexican display was specially remarkable for exhibits representing the ancient Aztec civilization.

Next to the Mexican, came the grand display of the United States, which occupied more than one-fourth of the floor-space in the main building, and was arranged on both sides of the main avenue. The exhibits comprised an infinity of articles, especially of those that tend to economize labour and time, and therefore, to reduce the cost of production. They showed how closely the States are treading on Great Britain's heels, in certain departments of manufacture; and with what right, they can now claim a foremost place amongst the great industrial and producing nations of the world. A large amount of space seems to have been allotted to each individual exhibitor, and unlike the British exhibits, a great deal of attention appears to have been devoted to the embellishment of the articles displayed; so that they had generally a more attractive appearance. They were in fact essentially exhibition goods, whereas the British exhibits, with few exceptions, seemed to be samples of the ordinary manufactures of the exhibitors. Never before has the multiplicity of American products, and the skill, ingenuity, and taste of her artizans, been so prominently exhibited ; and it appeared to me, that nothing but the high price of many of the commodities, will preclude them from successfully competing in the markets of the world. The quantity and bulk of the various exhibits called the attention to them of those visitors who might otherwise have passed them unnoticed ; such was the case with the cotton goods. There were numerous exhibits of plated-ware, and although they could not be compared to the splendid collection exhibited by Elkington in the British section ; they were yet noteworthy, for many improvements introduced. A great display was made of chandeliers and gasaliers of all description, some of most novel construction. The exhibits generally, comprised samples of all classes of manufactures, prominent amongst which, might be seen tweeds, prints, silk thread, paper, pianos, brass fittings, books, stationery, cutlery and edge tools, locks, fittings, crystal lustres, glassware, and a diversity of other articles. The tweeds though very good did not to my mind favourably compare with the Canadian. In glassware and more especially cut glassware the American exhibits did not come up by a long way to those of other countries.

Adjoining the United States Court, was that of Germany, which, compared with the large amount of space occupied by Great Britain, France, and the United States, was very small in size. It seemed to be the general opinion that the display in it, though very excellent in its way, was yet inadequate to give a correct idea of the high position occupied by the German Empire as a manufacturing country. The exhibits in this section, if not numerous, were however of a high degree of excellence, and could only have been the products of great skill, combined with artistic taste. They comprised samples of woollen fabrics, velveteens, beautiful glass-ware, porcelain-ware, brass musical instruments, pianos, surgical instruments, drugs, bronzes, gildings, and toys of ingenious construction, capable of raising the youthful heart into the seventh heaven of delight. All these exhibits displayed fine workmanship and finish, and a decided pre-eminence in many of the classes. The bronzes compared very favourably with those of their great rival France, but, representing as they for the most part did, the public men of the Empire ; there was not the same room for the display of those beautiful designs, in which the French excel, and by means of which, they have obtained such an acknowledged superiority in the fabrication of these goods.

Next to the German came the Austro-Hungarian Department, which contained a most attractive collection

of exquisite articles in Bohemian glass, many of which
might easily have been mistaken for Venetian manufacture ; stained glass for windows, beautifully enamelled
glassware, furniture, wood-carvings, amber goods, elegantly
carved meerschaum pipes, carpets, woollens, real and
imitation jewellery ; and a varied assortment of fancy
goods of all descriptions, forming a display, that was very
creditable to the great country that produced it.

The Russian Court, that was next in order, occupied
but a small amount of space, and that was not very well
filled.   The articles exhibited however, proved that
Russia is making great strides in its manufactures.
Amongst a host of other things, some silver-work in
repoussé was exhibited, of great taste and novelty, the
various designs being original, and displaying a distinctive
national character.

The Spanish Section, the one next to the Russian, was
entered from the transept, through a large and very elaborate entrance, constructed in three arches, in imitation of
pink granite, with bronze facings, draped with yellow and
crimson silk hangings.   This was surmounted by a design,
representing Spain drawing back a curtain and revealing
to the assembled nations the mountains and green valleys
of the American Continent.   Over one archway was a
picture of Columbus ; over the other one of Isabella III.
From the central arch hung a magnificent gothic candela-
brum of oxydised silver and brass, contributed by King

Alfonso. There were also other contributions from the Royal Collection, including carved woods inlaid with gold, splendid tapestries, china and porcelain vases, and other articles of rare beauty. An exquisitely carved sideboard by Forzano Bros. was a prominent exhibit in this section. The Azulejos,—tiles resembling Italian mosaics, —introduced into Spain by the Moors, were very much admired. The remaining exhibits were principally composed of damasks woven in arabesques, silks, magnificent specimens of ladies' hosiery, and numerous other equally beautiful *articles de luxe.*

The Egyptian Court had also a very ornate front to the main avenue, covered with representations of Egyptian architecture and bearing the appropriate inscription— "The oldest of the Nations sends Soudan,—the morning greeting to the youngest." In the manufactures of modern date, a great display was made of gold embroidered cloths, of richly decorated saddles, furniture inlaid with ivory and mother-of-pearl, samples of paper, splendid wood and metal carvings, woods, cotton and silk, both in the raw and manufactured states; together with some excellent specimens of printing in Arabic, Coptic, and Hieroglyphic characters.

The most interesting portion of the collection was, however, the fragments of the ancient monuments, and the antiques, that may have dated from before the pyramids. Amongst the many noteworthy objects in this

section, may be mentioned, three crystal lamps inlaid
with gold, from the great Mosque at Cairo, which are
supposed to have an almost fabulous value ; the art of
making them being amongst those things that exist no
longer.

Turkey made but a very poor show, doubtless in consequence of its political troubles, but an attempt was
made to cover the paucity of the display by the erection
of a gorgeous front.

Denmark made a very creditable display, principally of
works in terra-cotta, wood carvings, furniture, silver-work,
and furs.   This section well merited a few hours' close
inspection, as the various articles exhibited, were of excellent quality ; and the designs had an individuality about
them, very interesting to those accustomed to the better-
known manufactures of the great industrial nations.

Japan next came with a characteristic display of vases,
and other ornaments in elaborately carved bronze and
painted china ; of textile fabrics, and of numerous articles in
lacquerwork ; all representing the skill, and delicate mani-
pulation of this ingenious people, and their shrewdness as
men of business; for all their exhibits had a ticket
attached giving the price in plain figures.

The Chinese did not appear to have taken as much
trouble with their collection, as their neighbours, the
Japanese had done, and their display was consequently
not so good.   It consisted of articles characteristic of

Chinese manufactures, as silks, embroideries, painted screens, chinaware, ivory and other carvings, lacquerwork, papier-maché goods, articles made of soap-stone, &c.

Chili had a modest collection of minerals, native furs, skins, stuffed wild animals, and interesting relics of the Indian aborigines.

The small sections of the Argentine Republic, Peru, and the Orange Free State, were well stored with the various products of those countries.

We have now completed the circuit of the Main-building, but it is impossible, that a reader, from this cursory and brief description of its contents, could form a proper conception of the surpassing beauty and dazzling nature of this enormous exhibition. There are so many other buildings connected with this World's Fair to be visited, that we must at once proceed to the

MACHINERY HALL.

This structure might almost be called a continuation of the main building, having only been separated from it, by a small square, laid out in walks, and planted with ornamental shrubs and plants. The building consisted of a main hall 1,402 feet long and 360 feet wide ; together with an annexe on the south side. Its cost of construction was $700,000. The Main-building and Machinery Hall together, presented a frontage of three-quarters of a mile to the grand avenue.

The display in Machinery Hall was perhaps the most instructive and entertaining part of the whole exhibition ; for almost every machine used in the different manufactures was here to be seen in operation.

This section seemed to have the greatest attraction for visitors, and it was at all times difficult to approach the most interesting of the machines, on account of the crowds of people that always surrounded them. Standing in the centre of the vast collection, was the great Corliss Engine, that generated the motive power, by which all the machinery in the hall was set in motion.

Some idea of the magnitude of this enormous piece of machinery may be gathered from the fact, that it rose to a height of 40 feet above the floor ; that its gear-wheel was 30 feet in diameter, and 56 tons in weight ; and that it could work up to 1,400 horse power. In spite of its great size, this engine ran with great smoothness, and almost merited the rhapsodies into which American writers seemed to fall, when describing it.

Though the greater portion of the grand display in Machinery Hall belonged to the United States ; yet Great Britain, Canada, France, Germany, Belgium, Denmark, Sweden, and Italy were all more or less represented.

What seemed to strike every visitor to this section, was the high finish of the machines of American manufacture ; the various makers having gone to great expense in turning them out, resplendent in polished nickle and brass.

It will only be possible to enumerate briefly those machines that were the most interesting, and that attracted the most attention.

The Walter Press exhibited by the proprietors of the London Times newspaper, which was in operation, printing off copies of the New York Herald, was much admired; as were also several printing machines of American design and construction.

There were numerous exhibits of locomotives of great power and fine finish; enormous mining machinery; pumps of all kinds, the centrifugal especially being very excellent; blowers of such power, that they had to be cautiously approached; together with all the various machinery connected with wood-working, brick-making, barrel-making, ship-building, nail-making, watch-making, cork-cutting, shingle-splitting, &c. There were also improved fire-engines and escapes, dredges and apparatus for saving life at sea. In one machine exhibited, the attempt had been made to supersede steam as a motive power; and though the new agent is a trifle more costly, still it has so many other advantages to counterbalance that one disadvantage, that it may come into general use for many purposes. The motive power is gas, supplied from an ordinary burner, fixed near the bottom of the cylinder. Gas and air being admitted under the piston,

and exploded; the sudden expansion of the gas drives up
the piston rapidly, and by the time it has reached the end
of its stroke, the temperature falls, and condensation
produces a partial vacuum under the piston ; and the
downward stroke is effected by the pressure of the
external atmosphere. This is a German invention, and
the advantages claimed for it are, that no boiler or furnace
is requisite ; that it can be set in motion as soon as
wanted, and stopped almost instantaneously ; and that no
fuel is wasted when the machine comes to a standstill.
The Brayton hydro-carbon engine is an American
invention, in which a mixture of vapourized oil and water
is used, to produce the same effect, as in the one above
described.

The exhibitors of a safe almost circular in form, claimed
for it greater strength to resist fire and thieves, than the
safes of ordinary square shape. There was also one of
the chronometer safes shown ; these have two chrono-
meters affixed to the door, and in case one might go
wrong, both are set to the time at which it is desired the
door should be open. Until that time arrives, the door is
firmly closed, and cannot by any means be opened.

A very interesting machine was one for manufacturing
envelopes, which had only to be fed with paper and gum,
and would discharge the envelopes finished, and counted
into packets at the rate of 7,200 per hour. This machine
being open, the operation of envelope-making could be

seen in all its stages ; and there was generally an admiring crowd surrounding it.

There was an enormous display of weaving machinery, and the processes of manufacture of tweeds, carpets, cotton and woollen goods, could be seen in all the stages from the raw material to the finished product. In this department an exhibitor from Brazil displayed the process of silk manufacture, from winding the silk off the cocoon, to its being woven into beautiful articles of every-day use. Here also was shown a new invention for weaving horsehair, by which it is claimed the present cost of production is reduced one-third.

A number of rock-drilling machines for quarrying, and tunnelling purposes were always to be seen in operation, and the ease with which they pierced the hardest masses of stone was wonderful. In close proximity were machines for making gas in houses distant from a town, and which would be very useful for up-country districts in Australia. The well is filled with oil, and the machine wound up, like a clock ; when it will go on producing gas for some time without heat of any kind being used.

Nail-making machines were in great profusion. One that turned out 400 tacks per minute, was especially admired, as it was only requisite to feed it occasionally with a long iron rod. An india-rubber mill, where several men were engaged in manufacturing goloshes, was an interesting machine, showing every stage of manufacture, from the crude material, as it exudes from the tree, to the highly finished articles of commerce.

Some barrels, tubs, and buckets, made out of paper attracted much attention, and seemed very serviceable looking articles ; the barrels especially, though without hoops, and of straighter shape than is usual, were considered quite as strong as, and more economical than those made from hickory-wood. They are light in weight, cannot be broken, and hold fluids as perfectly as vessels of clay or metal.

A pretty little instrument, and one that may in time become a necessity in every counting-house, was the electrical pen, by the use of which the cost of lithography may be saved.

This pen is very simple in its construction, being made like a pencil-case, a needle taking the place of the lead ; this is connected with a small galvanic battery, which causes a wheel on the top of the pen to revolve and move the needle up and down, at the rate of several thousand strokes a minute. As this pen is moved along the paper, the needle punctures the writing, and forms what is called the " proof," which is then put in a small press, sold with the electrical pen, and by passing the roller over it, the requisite number of copies are made, which look as if they had been lithographed. For printing price-currents, circulars, &c., this instrument will, I imagine, in time come into use; its cost at present is £7, but this will, doubtless, be reduced when the demand for it increases. Another adaptation of electricity was to the sewing machine, which seemed to me likely also to come into general use. A parlour scroll-saw for cutting out wood and other materials, worked in the same manner as a sewing-machine, is an ingenious little machine. There were also shown stationary engines, worked by petroleum, at a cost of a few half-pence per hour. A very useful machine, and one that might advantageously be introduced into the colonies, was one for removing the burr from wool, without destroying the staple.

Another remarkably ingenious little instrument, was one intended to supersede the use of the pen. This was worked by pressing various keys, in the same manner as playing on a piano. A lady, who showed the working of this instrument, had attained such proficiency, that she could print at the rate of ninety words a minute.

A handy machine was one that weighs and packs parcels of flours, sugars, or spices : this would be very useful to grocers and other tradesmen. A patent cleat was exhibited, that renders unnecessary the holding of the corner of the sail, and that removes the danger that may arise from having the sheet belayed. This will prove a great boon to yachtsmen and others. The Covel saw-sharpener, that requires no supervision, is a machine that will most probably come into general use. The list of these useful machines might be continued at great length were it not that there is yet much more to be seen and commented upon.

Before leaving Machinery Hall, it may be well to know the opinion of American critics, on the relative merits of the English and American exhibits, in this section. England they acknowledge to excel the States in armour-plating; there being no rolling-mill plants in the latter, that could handle plates 21½ inches in thickness, such as were exhibited in the British section. England is also declared pre-eminent in " steam-hammers, road-steamers, steam-rollers, portable engines, steam-ploughs, hydraulic presses, marine engines, and mining engines of the Cornish pattern, sugar plant, and paper-making machinery." America is considered first in the manufacture of " harvesting, sewing, knitting, and hat-making machines; quartz and stone crushers, amalgamators, deep-well borers, weighing scales, breech-loading small arms, and special tools for the manufacture of articles made of many pieces, by the means of templets and gauges, the Enfield special machines made for the English armoury, and the Mauser machines for the armoury of Germany ; metallic fixed ammunition; watches made by special machines to scale and pattern, with interchangeable parts; wood-working tools and barrel machinery; wooden bridges, steam fire-engines, safes and safe locks, hay and cotton presses, sodawater machines, hotel and warehouse elevators, cheap good clocks, cast-iron car-wheels, web printing and folding machines, heating apparatus for houses, rubber goods, saws, and the extensive substitution of belting for gear in transmitting power." The two countries are supposed to divide honours in the following articles: "Planers, lathes, slotting and shaping machines, water-wheels, rotary pumps, blowers, locomotives, and steam gauges."

Having now briefly enumerated the principal objects of interest in Machinery Hall, we will proceed to another portion of the Exhibition, the

MEMORIAL HALL.

This building, used as an Art gallery, was erected by the State of Pennsylvania, at a cost of a million and a half dollars ; and was undoubtedly the finest of the Exhibition buildings. As it was intended to be a permanent addition to Fairmount Park, it is constructed throughout of stone, brick, and iron, in the modern Renaissance style, and is surmounted by a dome of iron and glass. It is 365 feet in length, 210 feet in width, and 150 feet high to the top of the dome, from which rises a colossal figure of Columbus. At the corners of the base of this dome stand four large figures, representing the four quarters of the globe ; whilst at the corners of the four pavilions are large cast-iron eagles, with outstretched wings. The main entrance consists of three arched doorways, opening into a hall, and between these, are clusters of columns, terminating in emblematical designs, descriptive of science and art. The doors are of iron, relieved by bronze panels, displaying the coat of arms of each State of the Union.

Though not by any means a model of architectural beauty, this building still forms a fitting memorial of the Centenary of American Independence.

Passing from the Main building, or Machinery Hall, to the Memorial Hall, which contained the exhibition of paintings and statuary, the change from the grand display in the two former, which quite overshadowed any previous exhibition, to the paucity of the Art collection exhibited in the latter, was very striking. It was not, that the collection was small, for it covered an immense wall space ; but there was a marked absence of works of high art. Great Britain contributed some fine water-colour paintings, and pictures by West, Lawrence, Gainsborough, Turner, Maclise, Millais, Holman Hunt, Leighton, and Frith. The latter's " Marriage of the Prince of Wales " was very much admired, and always had a large crowd around it. In the French department there were several specimens of gobelin tapestry, the designs on which included Maillard's "Penelope," Boucher's "Amynthe and Sylvie," and Boucher's "Fishing." There were also some paintings in this section of great merit, and the collection altogether was very creditable.

The Spanish contribution was small, but included two fine Murillos.

Germany had not a large collection, and the few paintings exhibited did not particularly challenge admiration. All the pictures had not been hung at the time of my visit, and this section would doubtless have assumed a more imposing appearance later on.

Italy and Austria both contributed large collections, but the same may be said of them, that has already been remarked of the other national collections, viz., that there was an absence of works of great merit.

The United States monopolized a large portion of the space, but the specimens of high art were few in number. The great picture, great on account of the enormous wall space it covered, filling up, as it did, the whole of one end of a large room, was Rothermel's "Battle of Gettysburg." This large picture did not, however, reflect much credit on American art. Two of the most prominent pictures in this section were West's "Christ Rejected by the Jewish People," and Pauwel's great allegorical painting of the "New Republic."

The Swedish collection was a very unassuming one, but contained one picture that attracted a great deal of attention, on account of its excellent delineation of a comical subject ; viz., two boys smoking for the first time. The fidelity with which the feeling of nausea was depicted on the boys' faces, was marvellous. The Australian art exhibits were not shown in the Memorial Hall, but were included in the general display made in the Main-building; in order to render the Australian show more effective.

AGRICULTURAL HALL.

Having "done" Memorial Hall and sufficiently admired the art treasures there exhibited, we proceed to the Agricultural Hall, which was built in Gothic form, the interior resembling that of a Cathedral; with a nave 820 feet in length, crossed by three transepts, each 540 feet long.

Some idea of the magnitude of the exhibition of agricultural products and implements in this Hall, may be formed, when it is remembered, that the exhibits covered a space over ten acres in extent; and that it was necessary to traverse three miles of walks to inspect them. It is only possible, without writing a volume on the subject, briefly to summarize what was to be seen in this department.

Great Britain and Ireland occupied only a small space, but their exhibits were pre-eminent in their respective classes. A great display was made of pickles, sauces, mustard, cocoa, chocolate, essences, and ærated waters; also, what seemed more properly to belong to this section; agricultural implements, edge tools, mill machinery, farm engines, road steamers, steam rollers, &c.

A machine much admired was one, that, with the aid

of two men, could sack, dry, and dress 100 bushels of wheat, or other grain, in one hour.   Another enables one man to separate fifty bushels of wheat or oats, from any other grain, in the same time.

The Canadians exhibited a splendid assortment of ploughs, mowers, threshing machines, churns, chaffcutters and numerous other agricultural machines, of excellent quality; and showed conclusively, that in this department of manufacture, they are quite able to hold their own against their neighbours in the United States, assisted, as the latter are supposed to be, by a protective tariff.   It may be remarked, that in the matter of agricultural implements, Canada and the United States monopolized nearly all the space in the Hall.   An excellent display was also made by Canada of wool, wheat, prepared skins, ales and preserved fruits.

Germany was great in wines and beer, in tobacco and cigars, in liqueurs and preserves; and the section also displayed a trophy composed of scythe blades and other farm tools.

Venezuela made a great display of such products as coffee, medicinal barks, samples of woods, starch, soap and candles, cordials and fruits.

Brazil which was in close proximity, exhibited within its handsome enclosure walls, samples of seeds and grasses, tobacco leaf, polished woods and liquid extracts. This section contained a little pavilion, constructed entirely of cotton fibre, in which were displayed samples of cotton seeds and plants.

Japan exhibited its teas, preserves, its numberless varieties of seeds, and curiously-made cane chairs.

The French exhibits had not all been arranged, and they did not seem to give promise that the collection would be a large one. As was to be expected the wines of Burgundy, Bordeaux, and Champagne occupied a prominent position.

The Netherlands succeeded in producing an impression of solid prosperity. The exhibits comprised grains, spirits, beer and liqueurs, seeds, various preserved foods and condiments, and numerous other exhibits, forming together a collection, that worthily represented its agricultural products.

Sweden and Norway had arranged their products in a very pleasing manner. They included samples of bottled ale, polished woods, grain, coffee, preserved fish, arrack, cod-liver oil ; together with sealskins and other furs. An attractive exhibit was the figure of a Laplander enveloped in costly furs sitting in a sledge, drawn by a reindeer. The Norwegian exhibits consisted principally of a number of aquaria, containing numerous specimens of fish, which proved a great source of attraction to crowds. There was in addition, a magnificent collection of nets, seines, rods and tackle ; together with all the implements connected with pisciculture.

Portugal was represented by a multiplicity of articles amongst which were wines, oils, wool, silk, cork, etc.

Italy was conspicuous with its wines, and oils, maccaroni and other food preparations; and though the collection generally was small, it was excellent of its kind.

Spain appeared to greater advantage in this department, than it had done in the Main-building; displaying here fine samples of grains, seeds, fruits, timbers, grasses, wines, oils, and other products of her teeming soil; illustrating how, with the blessings of peace, and the help of skilled agriculturists, it might attain the first position amongst European nations for products of the soil.

All these various national collections, interesting as they were, seemed only subsidiary to the colossal display of the United States, which occupied nearly three-fourths of all the space in the hall. Of the American collection it may be said that, whilst many of the States of the Union were but poorly represented, the whole exhibition collectively, was such, as to give a visitor a better idea of the vastness of American resources, than any other portion of the great World's Fair. The collection consisted of samples of the products of all the various climates and soils, from Maine to Louisiana, and from Florida to Oregon, the limits of the enormous territory in the possession of the United States; and in its range comprised the grain, wine, wool, and precious metals of California, and the raw and manufactured tobacco, sugar, rice, and cotton of the

Southern States. The State of Oregon displayed splendid specimens of merino and other wools, wheat and barley, oats, Indian corn, and linseed. The exhibits representing the agricultural products of most of the other States bore a strong family likeness to one another, and consisted for the most part of wheat, Indian corn, and other grains, woods, seeds, &c. Each State, however, exhibited in addition something peculiarly its own, thus Kentucky made a feature of its whiskies, and Minnesota was to the fore with a magnificent assortment of grains, representing its annual yield of thirty millions of bushels. The display of timbers from Delaware and Indiana was most imposing; the former having exhibited thirty-eight varieties, and the latter forty-three.

Exhibits of tinned fruits and vegetables were most numerous, and proved how rapidly this species of preserved food is coming into general consumption. The tinned fruits and vegetables only require warming, and thus a wholesome food is made available during the winter as well as the summer months.

One of the great features of this section was the mammoth grape-vine, from California, which in the last year or two showed signs of rapid decay, and was consequently dug up, and the trunk and principal limbs sent to the Centennial Show. The trunk is the thickness of a man's body, and the produce of the vine in one season was 7,500 bunches of grapes, of an estimated total weight of over five tons. This vine was supposed to be over fifty years old, and it is satisfactory to know that it has left offspring; one of which, though only sixteen years of age, is already a foot in diameter near the ground and covers with its branches an area of 10,000 feet.

In addition to these, were the thousand and one articles belonging to no State in particular, prominent amongst which were to be seen sugar-coated hams, little temples formed of confectionery, containing samples of the candies so dear to the American heart; biscuits, honey, chocolate, tinned fruits, starch, maizena, canned and dried fruits, tea and coffee, fertilizers of all kinds, hickory-wood barrels, and indiarubber in its raw and manufactured states.

This list gives but a very imperfect idea of the endless variety of articles exhibited in this department; and still no mention has been made of the agricultural machinery, which constituted the principal part of the display.

The most prominent of these exhibits comprised ploughs, harrows, winnowers, potato-diggers, horse-rakes, threshing-machines, hay-presses, stump-extractors, hay and chaff-cutters, lawn-mowers, fruit-drying machines, churns, reapers and binders, seed-sowers, maize-shellers, fruit and potato-parers, cherry-stoners, and fruit-preserving utensils; each of which merits a special description for its ingenious construction, and qualities as a labour-saver

15

and time-economizer; by means of which the cost of production is so much lessened.

There is yet so much to be seen, that, we must leave the Agricultural Hall, with this brief description of its contents, and proceed to the

HORTICULTURAL HALL

which was in close proximity. This is intended to be a permanent building, and is consequently more substantial and ornate, than the temporary structures. It is built in the Moorish style, with fine frescoes; and in its warmth of colouring, presents a great contrast to the sober hues of the Memorial Hall. It is constructed principally of iron and glass, is 383 feet in length, 193 feet in width, and consists of a large conservatory, and forcing-houses for young plants. The east and west entrances are approached by flights of marble steps, in the centre of each of which, stands a pretty little open Kiosk. The conservatory is ornamented with several fine fountains, and its flower-beds contain many thousand tulip and hyacinth bulbs, many of which were in bloom at the time of my visit, and presented a lovely appearance of rich colour. The horticultural collection comprised specimens from all the zones; and though scanty in its proportions, yet formed a most interesting portion of the Exhibition. The flora of Australia was poorly represented by a few palms, fig-trees, and eucalypti. The great feature of the whole collection was a specimen of Venus' fly-trap (*Dionaea*) which, with its sensitive tentacles, seizes its prey, and holds it fast, to be devoured at leisure. The fame of this rare carnivorous plant spread rapidly, and thousands visited the horticultural building to see it.

SUBSIDIARY BUILDINGS.

The five buildings above described formed the principal portion of the exhibition, but there were in addition numerous smaller structures in the grounds, used for special purposes.

One of these was the Women's Pavilion, a very pleasing and commodious building, erected by the Women of America at a cost of $30,000 collected by subscription. This building contained a nave and transept, each 192 feet long and was surmounted by a cupola. Its special purpose was to contain exhibits of female work only, and thus be a means of pointing out occupations of usefulness and profit, adapted to women. The collection in this building represented those pursuits, for which women are specially adapted, as sculpture, painting, literature, engraving, telegraphy, lithography, and education. Articles of female attire, with the exception of embroidery, lace and needlework, were excluded from exhibition. In this department, the Royal Society of Needlewomen, instituted, and presided over I believe,

by Her Majesty the Queen, exhibited some splendid
needlework, and tapestry. Those branches of education,
for which women are specially adapted, such as the
Kindergarten and object-teaching, were prominently
shown. A great attraction to this section were
looms, worked by women, employed in making carpets
and other delicate work. These were always surrounded
by an admiring crowd.

The Government Building was another of these
auxiliary structures, and in it were displayed exhibits
from the United States Treasury, War, Navy, and Interior
departments and from the Smithsonian Institute for the
purpose of "illustrating the administrative faculties of the
Government in time of peace, and its resources as a war
power; and thereby serving to demonstrate the nature
of our institutions, and their adaptation to the wants of
the people." Here was to be seen the operation of
making rifles and ammunition; here were cannon,
mounted and unmounted; gun carriages, shot and shell;
iron-plates and torpedoes; equipments for land and sea
service; models of bridges, forts and hospitals; models
of war vessels and a thousand other exhibits representing
the department of marine survey, and the weather signal,
and lighthouse services. The Smithsonian Institute
displayed illustrations of the geology, mineralogy, forestry
and natural history of the States.

Photographic Hall and Brewers' Hall were also subsidiary adjuncts to the Exhibition. The former was partitioned inside to form seven galleries, for the hanging of photographs, and the display of photographic appliances. The latter was erected by persons connected with the brewing interest, to show the process of manufacture, and the newest appliances used in the trade.

In addition to these were also many smaller buildings, erected by private enterprise, amongst which may be enumerated, a very complete glassworks; a pretty little pavilion, in which were exhibited Singer's sewing machines. Another contained coffins of so gorgeous a nature, that one felt almost tempted to die in order to occupy one of those grand receptacles. One was used as a bakery; another as a dairy; and still another contained exhibits of leather, boots and shoes. One of the most interesting of these smaller buildings, was a little wooden house, such as was used at the time of the Revolution, filled with revolutionary relics, each article of furniture possessing a history of its own. In this house also were several men and women dressed in the costumes of that period, and engaged in various occupations, the women for the most part spinning with the old distaff. Another interesting object was the Canadian log-house, composed of all the different kinds of wood grown in the Dominion.

There were several model school-houses erected in the grounds, one of the most perfect being the Swedish, which was admirably adapted for the purpose intended, that of a country school. A Turkish café had many visitors, as also a Tunisian café, where a woman dressed in the national garb, danced the scarf dance to the music (save the mark !) of three outlandish looking instruments.

Each State of the Union had its special Commissioner's house, and these were all different in design, and were for the most part attractive buildings.

The British Commissioners had a large building, repre- senting a squire's house of the sixteenth century, which with its tile roof, and big chimney stacks looked very well. It was furnished in the style of that period, and formed not the least interesting feature of the Exhibition.

The Japanese building was also very interesting, being a *fac-simile* of the houses used by the Japanese middle class. In its construction not a single nail had been used, all the material having been dove-tailed and mor- tised, and fastened with wooden pins.

The great Centennial Fountain, erected by the Catholic Total Abstinence Society, and intended to be a perma- nent ornament to the grounds, merits a few words. In the midst of a large circular basin, stands a pile of rocks, surmounted by a figure of Moses fifteen feet high, pointing upwards, to show the source of the miracle just per- formed, in bringing out the water from the rocks, at the stroke of his wand ; whilst the water gushing forth on all sides falls into the basin. Four arms stretch out in the form of a Maltese cross, terminating in four drinking fountains, each crowned by a statue nine feet high, representing—Commodore Barry, the father of the American Navy; Archbishop Carroll, the patriot priest of the Revolution; Chas. Carroll, of Carrollton, the Catholic signer of the Declaration of Independence; and Father Mathew, the Apostle of Temperance, who, with Chas. Duncombe, a Protestant minister, Richard Dowden, a Unitarian philanthropist, and William Martin, a Quaker, founded in Cork, in 1838, that society which was destined in a few years to count its converts by millions, and to spread its influence wherever the English language was spoken. These statues are all of marble; and round the basin are seven medallions, heads of Catholic soldiers and civilians who distinguished themselves during the Revolution.

About the grounds are fine statues of Dr. Wotherspoon, Wm. Penn, Columbus, and a huge granite monument to the American army, weighing, it is said, thirty tons. The bronze figure of Wm. Penn is thirty feet high, and is only exceeded in size by two bronze statues in the world. The statue erected by the brotherhood of the B'nai Berith, typifying religious liberty, is worthy of notice; the figures and pedestal are twenty feet high, and consist of a female figure representing American Liberty protecting a youth slightly draped, holding in one hand an urn containing the sacred flame.

The various exhibition buildings having been at some distance from one another, communication was effected by means of a narrow gauge railway, which ran round the grounds, stopping at the principal buildings.

Before closing this chapter, I may mention that the preparations made for the accommodation of the anticipated great influx of visitors, were very complete, and there was consequently no difficulty in obtaining apartments. In the hotel where I lived, the Globe,—a temporary building erected just outside the Exhibition grounds,—there was accommodation for 3,000 guests; and during the time of my stay, there were never more than 1,500 in the house. There were also numerous other large temporary hotels near the grounds, at all of which accommodation could be had at the usual American hotel rates. Private apartments could also be obtained in the city at very moderate cost.

CHAPTER XV.

BALTIMORE, ANNAPOLIS, AND WASHINGTON.

JOURNEY to Baltimore—Description of the City—Monuments—
Public Edifices—Commerce—Annapolis—Senate Chamber—
Naval College—District of Columbia—Decentralization—The
City of Washington—Its Appearance—The Capitol—Ameri-
can Speakers compared with English — White House—
Treasury—Patent Office—Other Public Buildings—National
Memorial—Smithsonian Institute – Corcoran Gallery of Art—
Howard University.

AFTER having spent a couple of weeks in Philadelphia,
I was soon conveyed by rail to Baltimore, the chief city
of Maryland, and in population and commerce, one of
the principal in the United States. The scenery on the
route from Philadelphia to Baltimore ·is uninteresting,
although numerous well-cultivated farms, and many
thriving towns are passed. Amongst the latter may be
enumerated Chester, the oldest town in Pennsylvania, it
having been settled by some Swedes in 1643 ; Wilmington, the capital of the State of Delaware with a population
of over 30,000, and the seat of many important industries ;
and Newark, a pretty little place, containing the well-
known Delaware College. Shortly after leaving the latter
place, the train crosses the celebrated Mason and
Dixon's line, so long the boundary between the Northern

and Southern States.   At Havre de Grace we crossed the
beautiful Susquehanna River, with its numerous pretty
islands, on a bridge nearly a mile in length.   This has
been erected in place of the one destroyed by the
Confederate soldiers in the late Civil War, for the
purpose of cutting off the Federal communication with
Washington.

Baltimore is situated on the Patapsco River, about
fourteen miles from the entrance to Chesapeake Bay ; its
harbour is very capacious, and consists of an inner and
outer basin, protected by Fort McHenry.   The city is
very much like a large English provincial town in
appearance, and cannot by any means be called pretty ;
the streets being narrow, and closely built upon ; but it is
one of the cleanest towns in America.   The inhabitants,
who pride themselves upon being the handsomest people
in the States, are quiet and orderly ; and the town
generally has a more homely appearance than is usually
the case with American cities.   Altogether I would as
soon live in Baltimore, as in any town in the United States.

Baltimore is sometimes called the " Monumental City ;"
why,—it is difficult to imagine ; because it only possesses
three Monuments, which are however very fine.   The
Washington Memorial is a Doric shaft 176 feet in height,
resting on a huge pedestal, and supporting a colossal
statue of Washington ; it forms a very imposing landmark.
Battle Monument, erected to the memory of those, who

fell in defending the city in 1814, when attacked by the British, is also a fine column in the form of a Roman Fasces, surmounted by a female figure representing Baltimore, the base being in the form of an Egyptian temple. There is also a fine statue erected in honor of Thomas Wildey, the founder in the States of the order of Odd Fellows.

The principal edifice in the city is the new City Hall, one of the finest municipal buildings in the country. It is built entirely of white marble, in the composite style, surmounted by an immense dome; but it stands in a narrow street, where its fine proportions are not seen to advantage. The Exchange is a fine building, having on its two façades colonnades of six Ionic columns, the shafts of which are single blocks of marble of admirable workmanship. It is crowned by a large cupola, beautifully frescoed in the interior, and contains a fine reading-room. In this building are the Post Office, the Custom House and the Merchants' Bank. The Baltimore Athenæum is a fine institution, with a library of 26,000 volumes. The Peabody Institute is a fine white marble building erected by the late George Peabody the well-known philanthropist; it contains a free library of 56,000 books, and in connection with it, is an Academy of Arts.

The Catholic Cathedral is a stately edifice of granite, and contains two excellent paintings,—"the Descent from the Cross," presented by Louis XVI., and—"St. Louis burying his officers and soldiers slain before Tunis," the gift of Charles X. There are numerous other fine church edifices, foremost amongst which may be enumerated Grace Church, Christ Church, Emanuel Church, St. Paul's and St. Peter's, all of which are Episcopal. The First Presbyterian is an elaborate building and the Unitarian Church is also a handsome and very unique structure.

There are many Educational and Charitable institutions in the city, some of them possessing fine buildings, and being of a very high order; in fact, in this respect, Baltimore is better provided than most of the other large American towns.

The population of Baltimore is over 300,000, and its commerce is very important; it being a large port of export to Europe for tobacco, cotton, petroleum, bacon, dairy produce, &c. It also contains large smelting works, and is the centre from which the rich copper mines of Lake Superior are worked. Its industries consist of ironworks, rolling-mills, nail factories, locomotive works, and cotton factories ; and its tanning trade is also large, it being computed, that half a million of hides are annually tanned and exported to New England. The tinning of oysters, fruits, and vegetables, is carried on very extensively ; the annual value of this industry being estimated at a million sterling.

I availed myself of an opportunity that offered to proceed down the Patapsco River, and Chesapeake Bay, to
visit Annapolis the capital of Maryland. The view of
Baltimore from the river is very picturesque, and the trip
down the Bay is most pleasant. Whilst at Annapolis,
I visited the Senate Chamber in the State House, memorable for being the room, in which Washington resigned
his commission as Commander-in-Chief of the forces, into
the hands of Congress ; after the objects of the Revolution
had been attained. It contains a fine painting commemorating this scene ; and Washington's address, and the
reply of Congress, are also to be seen. The principal
Naval College of the United States is at Annapolis ; the
buildings, situated in extensive grounds, seem to be
excellently well adapted for the purpose, and the internal
arrangements are said to be very good. In addition to
an ordinary wooden training-ship, there is an iron Monitor
in connection with the College ; so that the cadets are
practically trained in the management of that class of war
vessel.

Returning to Baltimore, I took train for Washington,
the political metropolis of the Republic. Washington
is situated on the north bank of the Potomac River,
within an area of ten square miles, reserved for the Capital,
in the District of Columbia. This Federal District of
Columbia contains about sixty square miles,—its chief cities
being Washington, the capital of the Union, and George-

town. The object of making a Federal district, in which to place the capital of the country, was, to prevent any one State from exercising undue influence, by having the metropolis within its jurisdiction. The old law-makers of the Republic, in its early days, seem to have regarded this point as very essential to the common weal ; and jealously watched that no one State gained an ascendancy over the others, in the Councils of the country. They had also in view the dangers of centralization, and in every case, second-rate towns were made the political capitals of the different States ; so that, at the present time, it is seldom the capital of the State is at the same time its chief city.

It is a singular anomaly, that Washington, the centre as it were of a purely democratic republic, has neither part nor lot in the authority to which it is subject. It sends no member to Congress, being in this respect less favourably situated than the Territories, which do return delegates to speak, if not to vote. The District of Columbia, in short, has taxation without representation.

The city of Washington, originally designed by A. Ellicott, during the presidency of Washington, who by the way wished it to be called " The Federal City," was laid out on a very grand scale, as it was anticipated that it would become an immense metropolis. This has, however, not proved the case, as the city has no commercial importance : it has consequently a ridiculously straggling appearance, and in it, palaces alternate with buildings that may comparatively be termed shanties. It is built on the rectangular parallelogrammic plan so common in America. The streets are all wide, and shaded by fine trees, and are divided into numbered streets, lettered streets, and avenues: thus, First, Second, and Third streets are crossed at right angles by A, B, and C streets; whilst the avenues, named after the different States of the Union, form the main arteries of the city. The principal of these is Pennsylvania Avenue, especially that portion which extends from the White House to the Capitol, and in it are situated most of the public buildings; whilst the best retail establishments are in Seventh street. The city contains a population of about 110,000 inhabitants; but this number is much increased during the sessions of Congress, when it swarms with political agents, needy office-seekers, and general hangers-on.

The great feature however, of Washington, is its public buildings, foremost among which, stands the grand Capitol, one of the largest, and of its kind perhaps the finest edifice in the world. It crowns the summit of Capitol Hill, and consists of a main or central building, 352 feet long, and 121 feet deep, and two wings each 238 feet by 140 feet, covering together 3½ acres of ground. It may be said to be three distinct Grecian temples, each having a rich Corinthian portico, the centre one being surmounted by an immense dome, which altogether dwarfs its noble proportions, and which though painted to resemble marble, is constructed of iron. The central building is constructed of light yellow freestone painted white ; the two extensions are of pure white marble, and the general appearance is severely classic, although it has a great fault, inasmuch as it fronts up hill, and its principal façade is turned away from the City. It stands within some thirty acres of grounds, beautifully laid out, and adorned with statuary ; and in front of the building stands Greenough's colossal statue of Washington. On either side of the entrance are large figures of Peace and War ; and over the door-way is a basso-relievo representing Fame and Peace, in the act of crowning Washington with laurel. The Rotunda is the most striking feature of the interior of the Capitol ; it contains eight large, though they cannot be called fine pictures, illustrating American history, for they possess little artistic merit. They consist of the " Declaration of Independence," " the Surrender of General Burgoyne," " the Surrender of Lord Cornwallis," " Washington resigning his Commission," " the Landing of Columbus," " the Discovery of the Mississippi by De Soto," " the Baptism of Pocahontas," and " the Embarka-tion of the Pilgrim Fathers." There are also alto-relievos representing " Penn's Treaty with the Indians," " the Landing of the Pilgrims at Plymouth," " the Conflict of

Daniel Boone with the Indians," and "the rescue of Captain John Smith by Pocahontas." Over this rotunda, the dome rises to a height of 250 feet, and is beautifully frescoed with sixty-three figures, so large, that they look life-size when viewed from the floor. The design is the figure of Washington, sitting between the Goddesses of Liberty and Victory; below are the original thirteen States holding up a banner inscribed with the national motto *" E pluribus unum,"* and surrounded by six allegorical groups representing War, Agriculture, Mechanics, Commerce, the Navy, and Science. In the latter group Franklin, Fulton and Walter occupy prominent positions.

The old Hall of Representatives, now used as a National Hall of Statuary, is semi-circular in form; the entablature being supported upon twenty-four columns, and the ceiling painted in imitation of that of the Parthenon at Rome. Over one of the entrances, is a fine figure of Liberty; over the other, a statue representing History in a winged car, the wheel of which forms a clock. The Chamber of Representatives is a magnificent hall, the ceiling being of ironwork, with forty-five stained-glass panels, on which are painted the arms of the States. On either side of the marble desk of the Speaker are full-length portraits of Washington, and Lafayette. The accommodation for the public is excellent, but like the House of Commons at Westminster, the

16

acoustic properties are not good. The Speaker's room, immediately behind his desk, is a richly-decorated apartment.

The Senate Chamber is somewhat smaller than the Hall of Representatives ; it has spacious galleries for visitors, reached by fine marble staircases, which are amongst the most striking of the internal architectural features of the Capitol. Other fine chambers are the President's, and Vice-President's rooms ; the Reception-room, and the Senate Post-Office ; and especially the Marble-room, which is particularly chaste and rich in its decorations.

The Chamber occupied by the Supreme Court is very fine, and contains some beautiful marble Ionic columns. The Judges when presiding wear black gowns, but barristers address the Court in their ordinary dress. The Law Library, in connection with the Court, contains 30,000 volumes. The Library of Congress, the largest in America, numbers 300,000 volumes.

I was fortunate in witnessing a sitting of Congress, and must say that although I heard some of the leading Senators and Congressmen speak, I cannot say I was particularly impressed by their eloquence. They do not appear to me to speak in the quiet undemonstrative manner, that carries with it weight and conviction, but have a jerky style, use very strong expressions to describe very ordinary events, and in consequence of the habit that generally prevails of raising the voice at the end of a sentence, have a sing-song manner that is not pleasant. They seem however to possess a ready flow of words, such as goes by the generic term of " gift of the gab." I would say, that a comparison between the British Parliament, and the Congress of the United States, would show that, although the former is pre-eminent in the number of first-class speakers; the latter would have a better average of general debating power. There seems, however, to be a want of dignity in the deliberations of the two Houses of Congress, that must be very apparent to a stranger. The lobbies just outside the Chambers, in which the two branches of the Legislature are sitting, with open doors, are noisy with the sound of people walking about, speaking in loud and excited tones, and lobbying members as they leave the Chamber.

The White House, the residence of the President, is built of freestone, and painted white; and though what would be considered a fine mansion for a private gentleman, is too unpretentious in my opinion to be the state residence of the Head of the Republic. The front is very plain, relieved only by a portico supported on ten Ionic columns. The state-rooms are handsome, but they are not very commodious, and must be inconveniently crowded when state receptions are held in them.

The Treasury is a colossal building of white granite, and is the best adapted for the purpose, perhaps, of any

in the world. The east front has an unbroken Ionic
colonnade 342 feet in length, modelled after that of the
Temple of Minerva, at Athens. Extensive additions to
the building have been made in harmony with the general design, and it is altogether 582 feet long and 300 feet
wide, and cost six millions of dollars ; the interior is
ornamented with combinations of different kinds of
marble, and is very chaste in appearance.

The Patent Office, which also contains the Department of the Interior, is undoubtedly the finest of all the
public buildings in Washington. It is built of marble,
in severe and massive Doric style, with a grand portico
on each of its four façades; that on F street.is reached
by a broad flight of steps, and consists of sixteen Doric
columns of immense size, supporting a classic pediment.
The model-room consists of four large halls, with a
united length of nearly a quarter of a mile, and is filled
with cases containing about 120,000 models, representing
inventions in every branch of mechanical art, for which
patents have been granted. A large hall in this building
contains a collection of revolutionary relics, amongst
which are Benjamin Franklin's printing-press, and many
personal souvenirs of Washington.

The enormous building for the State, War, and Navy
Departments is now nearly completed, the State Department having already moved in. It is built of granite,
and exceeds in size even the great Treasury building,

being 567 feet long, 342 feet wide, four storeys high, with lofty mansard roof.

The General Post Office is an imposing edifice of . white marble, in the modern Corinthian style, and in general harmony with the other public buildings. It seems strange, however, that such a large building should be requisite for a city without any commercial importance, even though it be the metropolis.

The Department of Agriculture is a handsome brick building in the Renaissance style, three storeys high with a mansard roof, and contains a library, a museum of agriculture, a herbarium with 25,000 varieties of plants, and an entomological museum. The flower garden in front of the main building is a fine sight when in bloom. Connected with this Department are also an arboretum, experimental gardens and plant houses.

Washington possesses fine equestrian statues of Generals Washington, Scott and Jackson; and in addition to these, there has been for several years in course of construction, a National Memorial to Washington, which in its less than half finished state, is an ugly object at best, and spoils many otherwise pretty views. It is intended to represent an Egyptian obelisk, but as it is being built of small stones will never be handsome, if it ever be completed. Of this however, there seems much doubt, as there is a difficulty in collecting the requisite funds; and it is now suggested to demolish the portion already erected, and with the material to construct an Arc de Triomphe. This would certainly be advisable in the interests of good taste, but would be a humiliating end to the National Memorial to the national hero.

The Smithsonian Institute is a beautiful building of red sandstone in the Norman style, with many pretty towers. It was founded by James Smithson, an Englishman, in 1786, who bequeathed his large fortune for the purpose of erecting a building at Washington to be called the Smithsonian Institute, for "the increase and diffusion of knowledge amongst men." Regarding this bequest as a benefit for mankind in general, Congress passed an act for the erection of a suitable building, to contain a library, museum, art gallery, and a lecture hall; leaving it discretionary with the trustees, to use the remainder of the funds in any manner, that would carry out the wishes of the founder. The Institute now contains a fine museum of natural history, arranged in a series of halls, and fine ethnological, mineralogical, and metallurgical collections. It is situated in very fine and extensive grounds; and a world-wide reputation has been earned by its Transactions, which are annually published, and distributed amongst kindred institutions, and which have proved of great benefit to the scientific world, and consequently to mankind generally, as was originally intended by the beneficent founder.

The Corcoran Gallery of Art is a large building of brick and brown stone, in the Renaissance style, and contains some fine paintings and statuary, and collections of bric-a-brac and majolica-ware. Amongst the statuary, Powers' "Greek Slave" is a prominent object.

Howard University is a noble institution, founded for the education of youth without regard to sex or colour; it occupies a large brick building, painted white, surmounted by a fine tower, and gives instruction at present to 700 students, all negroes.

CHAPTER XVI.

CINCINNATI, LOUISVILLE, AND ST. LOUIS.

JOURNEY from Baltimore—Scenery—Harper's Ferry—Journey Resumed—Arrival at Cincinnati—Its Position—General Appearance—Tyler—Davidson Fountain—Public Edifices—Eden Park—Spring-grove Cemetery—Trip down the Ohio—Description of Louisville—Commerce—Unrivalled Position of St. Louis—Its Progress—Appearance—Streets and Buildings—Mississippi Bridge—Characteristics of Western Men.

RETURNING to Baltimore, I soon found myself ensconced in a comfortable carriage of the Baltimore and Ohio Railway Company, *en route* for Cincinnati. This is a journey of close upon 600 miles, and occupies twenty-four hours. The road passes through beautiful scenery, and many places in its vicinity have been the scene of exciting events during the Civil War ; so that instead of being monotonous, this journey is a most pleasurable one. After leaving the city, we crossed the Carrollton Viaduct, a splendid granite bridge spanning Gwinn's Falls ; and soon entered the " Deep Cut," a cutting seventy-six feet deep, and nearly half a mile long, which formed one of the greatest difficulties in the construction of the road. The route continued interesting until Washington Junction was reached, where the train entered the gorge, through which the Patapsco river flows. On entering this defile, we obtained a fine view of the Thomas Viaduct, a fine granite bridge nearly 700 feet long, resting on eight elliptic arches and crossing the river at a height of sixty feet above the water-level. We soon arrived at a little place called Ellicott's Mills, situated in a rocky gorge, through which the Patapsco tumbles in a most excited manner. After leaving this picturesque village, we passed many striking bits of scenery, and after twice crossing the river on fine viaducts, we arrived at Frederick Junction. This place had been the theatre of a most sanguinary struggle between Federal and Confederate soldiers, which resulted in the defeat of the former. The road now passed over a fine open country extending to the Catoctin Mountains, a continuation of the Blue Ridge range; and we obtained a fine view of one of the most noted features on this route, called Point of Rocks. This is a lofty promontory formed by the Catoctin Mountains, round the base of which, the Potomac flows, and which completely blocks up the pass; but a long tunnel cut through the solid rock enabled us to pass this impediment to our further progress. For nearly three miles before reaching Harper's Ferry, the track ran through a romantic defile in the mountains, the rocky side of which rises to a great height, looking like a great wall of stone. Harper's Ferry is most beautifully situated at the base of a high hill, at the confluence of the Shenandoah and Potomac rivers. The scenery in the neighbourhood is grand; the united waters of the two rivers flowing between the Maryland Heights on the one side, and the Bolivar Heights on the other; and I was very much tempted to break the journey here, and spend a day or two amidst this picturesque scenery. At Harper's Ferry, we crossed the Potomac on a fine iron bridge, and proceeded through the ravine of Elk Branch, until we entered upon a fine open undulating country, which extended to the town of Martinsburg. The route here became uninteresting, until the Potomac was again reached, at a point opposite the ruins of Fort Frederick, after passing which, the road swept round the base of a mountain, past a remarkable insulated hill called "Round Top." Here we commenced the ascent of the mountains, during which we obtained many fine views of the surrounding country, and passing through the celebrated Doe Gully tunnel 1,200 feet in length, and Paw-Paw tunnel, we continued our way for some distance through rugged and imposing scenery; and after again crossing the Potomac we reached the town of Cumberland. This town lies in an amphitheatre surrounded by the mountains, and the approach to it is most striking; it is in point of population and commerce the second town in importance, in the State of Maryland. The scenery after leaving Cumberland continued very picturesque, and at the place where the Potomac was crossed, and the train passed from Maryland into Virginia, the views up and down the river were very fine. We soon commenced the ascent of the Alleghany mountains, and in a short time reached Altamont, situated on the extreme summit of the Range. Leaving Altamont, we passed through beautiful natural meadows, locally called "glades," watered by numerous streams ; and commenced to descend the mountains through big excavations and tunnels, until we reached the valley of the Cheat river, passing which, we again descended through Kingwood tunnel, which is 4,000 feet long, and at Grafton left the mountains behind us. From Grafton the country was well-wooded, but uninteresting, until we arrived at Parkersville, where we crossed the river Ohio, on a grand bridge considerably over a mile in length, which spans the river upon six arches, the approaches to it resting upon no less than forty-three arches. We soon reached Athens, a pretty little town on the Hocking River. Athens contains the Ohio University, the oldest seat of learning in the State. The next place of importance was Chillicothe, a town situated on a fine plain, through which the Scioto River runs ; and which, at one time, was the capital of the State of Ohio, and is still a thriving, and very pretty town. Thence to Cincinnati, there was nothing noteworthy on the route, though the country we passed through, seemed well cultivated.

Cincinnati, founded in 1788, and consequently a little under a century old, is certainly the most picturesque of the great cities of America; it is situated on the Ohio River, at this point about twice as broad as the Thames at Hungerford Bridge; its waters, yellow as those of the Tiber, here separating the State of Ohio from the adjoining one of Kentucky. On the Kentucky side of the river opposite Cincinnati, and connected with it by a magnificent iron suspension bridge 2,250 feet in length, the span between the towers being over 1,000 feet, are the cities of Covington and Newport. The view from the surrounding hills, of the three cities and the winding river, is very fine. Cincinnati is built on two terraces, rising one above the other, at a good elevation above the river; and being entirely surrounded by hills on three sides, thus lies in an amphitheatre; and this position gives it a picturesque variety of scenery, the want of which is so apparent in the majority of the large American cities. It is regularly laid out, the streets being broad, well paved, and crossing one another at right angles; although they are not as clean as they might be. This, by the way, is a failing common to all the cities of the Western States.

The business portion of Cincinnati is compactly built, the buildings being for the most part of a dark freestone, which gives them a substantial appearance; the private residences are situated on the upper terrace, and on the hills in rear of the city.

Fourth street is the fashionable promenade, and contains the finest shops. In Pearl street are the principal wholesale warehouses; and these being generally uniform high stone buildings, have a very imposing appearance. Third street contains the banks and insurance offices. There are, besides, many fine streets with beautiful private residences.

A prominent feature in the city, is the beautiful Tyler-Davidson Fountain, erected at a cost, it is said, of £40,000, by Mr. Davidson, and by him presented to the citizens. It is of bronze, of exquisite workmanship; but it was cast in Munich, and though creditable alike in design and execution, it certainly cannot be regarded as a specimen of American art.

The population of Cincinnati now approaches a quarter of a million of inhabitants, a third of which are either German or of German parentage, and occupy a portion of the city north of the Miami Canal, or the Rhine as they have re-named it. Crossing this canal, one finds oneself in seemingly quite a different country; no other language but German being heard, and the general appearance of the houses, and more especially of the numerous beer gardens, reminding one forcibly of the Fatherland.

The Government buildings do not call for special mention; the County Court House being the only one with any pretensions to architectural design, but there are several very excellent charitable institutions, amongst which the Cincinnati Hospital and the Longview Asylum for the Insane are prominent. The city too is particularly well provided with first-class educational establishments, many of them of a very high order.

The principal church edifice in the city, is the Roman Catholic Cathedral of St. Peter, a fine building in pure Grecian style with a high spire, and a portico supported upon ten columns. The altar-piece " St. Peter Delivered," by Murillo, is one of the finest paintings in America. The Episcopal Churches are plain and unpretentious; but the First Presbyterian is a fine building, with a huge tower surmounted by a spire 270 feet high, terminating in a gilded hand, the finger pointing upwards. The Hebrew Synagogue is a profusely ornamented edifice in the Moresque style, with the most beautifully decorated interior in the city. The Hebrew Temple is a Gothic edifice with double spires, and its interior is very beautifully frescoed.

Eden Park is a fine piece of ground, containing 216 acres; it is situated on a breezy hill, which commands fine views of the city and surrounding country; is well laid out, and contains the two reservoirs that supply the city with water, and which very much resemble natural lakes. Burnett Woods is a fine tract of forest-land, 170 acres in extent, situated at a short distance from the city, and forms a pleasantly-shaded recreation ground, much resorted to by the citizens of Cincinnati in the hot weather. There are also numerous small parks or " greens " scattered about the city, which are a great boon to the inhabitants, and in addition to beautifying the town, must exert a beneficial effect on the general health.

One of the lions of Cincinnati, is Spring-grove Cemetery, distant about three miles from the city. This is approached by a fine avenue, and consists of some 600 acres, well wooded, and very picturesque. The entrance is very beautiful, being in the Norman-Gothic style, and the Cemetery contains a fine bronze statue of a soldier, cast in Munich, and erected in honour of the Ohio volunteers, who fell in the Civil War.

Cincinnati is the seat of many important industries; the manufactures of the city are valued at £30,000,000 per annum, and consist principally of iron, boots and shoes, machinery, steamboats, furniture, beer and whisky. Pork-packing, however, is the great industry; and in this branch it ranks directly after Chicago.

From Cincinnati I proceeded down the Ohio to Louisville. The view from the steamer on starting was very fine; on the one side the city rose, terrace above terrace, towards the hill-tops, which, covered with villa residences, and vineyards, formed a beautiful semi-circular background; whilst on the other side lay the twin cities of Covington and Newport, nestling at the foot of the Kentucky Hills. The river scenery is uninteresting;

towns being at great distances apart, and separated by
large tracts of virgin woodland, or plains, which have a
great sameness, and tend to make the trip monotonous.
The land on the upper Ohio, above Cincinnati, is said to
be under cultivation, and to be of a more diversified
character, with pretty homely scenery. Below Cincinnati,
the river, which had hitherto formed the boundary
between the States of Ohio, and Kentucky, makes a
sudden bend, and becomes the dividing line between the
States of Indiana and Kentucky.

Lawrenceburg and Aurora passed by the steamer are
flourishing little towns in Indiana, and have a rather
considerable shipping trade. Big-Bone Lick on the
Kentucky side, is so called from a quantity of mastodon
bones having here been found. Carrollton stands at the
junction with the Ohio of the Kentucky river, which is
navigable for about 200 miles, and possesses very picturesque scenery. Madison one of the principal cities in
Indiana presents a very imposing appearance from the
river ; it is well built and is a place of importance, having
a large commerce. Approaching Louisville the view
becomes very fine ; the river, here about a mile wide, is
crossed by an immense bridge, which connects the
Northern and Southern railway systems ; whilst the view
of Louisville on the one side, and of Jeffersonville on the
other, is really imposing.

Louisville, the chief city of Kentucky, is situated on a

plain surrounded by hills, on the Ohio, near the junction with that river of the Bear-Grass creek. Opposite the city are the falls of the Ohio, which are very picturesque, being a succession of small cataracts, extending right across the river, but which disappear when the water is high. To prevent the navigation of the Ohio from being impeded when its waters are low a canal has been constructed round the falls, at great cost, to a place called Shippingport.

The city covers an area of thirteen square miles, with a frontage to the river of three miles, and is laid out with great regularity; the streets being straight, wide, well-paved, and shaded by fine trees, but like those of Cincinnati they might with advantage be kept cleaner. The first settlement on the present site of Louisville was made in 1778, and the town was established in 1780, and named after Louis XVI., King of France, who was then assisting the States in their struggle for independence. It now contains over 100,000 inhabitants.

The public buildings of Louisville are plain and substantial; but have no claim to beauty of design. The charitable institutions are numerous in proportion to the population, and the educational establishments and libraries are many and excellent; the new coloured Normal school for the instruction of negroes being one of the finest schools of the kind in the country.

The private residences in the city have generally nice

17

lawns and gardens in front, and in this respect form a
pleasing contrast to those in the other large towns, where
the absence of private gardens is very noticeable.

On the Indiana side of the river, opposite Louisville, is
the town of Jeffersonville, reached by an iron bridge here
erected over the Ohio. This bridge or viaduct, the pride of
Louisville, is a mile in length, supported upon twenty-four
piers, and cost £400,000 ; but unlike the beautiful bridges
at Cincinnati and St. Louis, it is not pleasing to the eye.

The commerce of Louisville is very considerable, it being
one of the largest leaf-tobacco markets in the country,
and also a great emporium for provisions and live stock.
Pork-packing and ham-curing are large industries, and
it is the distributing market for the wretched Kentucky
whiskies, the consumption of which is enormous. It is
also the centre of several important manufactures, the
principal being leather, cement, furniture, and agricultural
implements. The casting of iron, water and gas pipes is
also a large industry. The total annual value of the trade
of the city is estimated at £50,000,000.

After spending a couple of days very profitably at
Louisville, I took the cars of the Ohio and Mississippi
Railway, and after a rather uninteresting journey arrived
at St. Louis.

St. Louis, the metropolis of the West, and the chief
city of the State of Missouri, is situated on the Mississippi
River, about twenty miles below its confluence with the

Missouri, and 170 miles above its junction with the Ohio. Its position is unrivalled; situated on a river, that, with its tributaries commands half the traffic of the whole country; in close proximity to enormous tracts of the best agricultural land, and to almost boundless forests of fine timber; with inexhaustible resources in its coal and iron deposits; there would appear to be no limit to the future greatness of this city. Its progress hitherto has been so rapid, equalled only by that of its great rival Chicago, that it seems destined in time to become, the greatest of American cities.

The first settlement made on the present site of St. Louis, was in 1764, by Pierre Laclede and others, who, under the title of the Louisiana Fur Company, received from the Governor of Louisiana, then a French colony, a grant of land, and permission to establish trading-posts on the Mississippi; and here the principal post called St. Louis was settled. In 1813, when Louisiana was ceded to the United States, that portion of it situated above the 33rd degree of latitude, was erected into the Missouri Territory, and eventually received as a State into the Union. The growth of the city has been marvellous; in 1811 the population amounted to 1,400 souls, and it is now supposed to contain nearly half a million of inhabitants.

St. Louis is built on three terraces, rising one above the other from the water's edge, and is, in spite of its smoke, and dust, one of the finest cities in America. It extends along the river for a distance of eleven miles, and covers an area of 21 square miles; the business portion being densely built, and the whole river-front embanked, and forming a Levee, as wharves are called in America. Built for the most part of stone, the city has a very substantial appearance, and is laid out with great regularity, the streets near the river running parallel with it and further back, being at right angles to those that cross them. The streets running north and south, or parallel with the river, are numbered First, Second, Third, etc.; those extending from east to west have mostly pomological names ; and the houses, being all numbered on the Philadelphia plan, it is easy to calculate one's distance from Market street, the centre of the city, or from the Levee.

Front street extending the whole length of the Levee, contains fine blocks of warehouses, and together with First and Second streets, is the centre of the wholesale trade of the city. Fourth street is the principal promenade, and in it are the finest shops ; whilst the private residences are in the avenues, and in Pine, Olive, and Locust streets.

St. Louis possesses some fine public buildings. The Court House is built of limestone in form of a Greek cross, with fine Doric porticoes, and surmounted by a large cupola, from the top of which a grand view of the city, river, and surrounding country is obtained. The Four Courts building, in which, as its name implies, the different Courts are held, is a beautiful freestone edifice, very ornate, having in its rear a jail, . constructed of iron, semi-circular in form, so arranged, that all the cells can be overlooked at the same time by a single warder. A new building is in course of erection, to be occupied as a Post Office and Custom House, and will, when completed be very fine, and a great improvement on the present inconvenient building; as will also the new Exchange now being built. Other fine edifices are the Masonic Temple, the City Hall, the St. Louis Life Insurance building, and the Republican Newspaper building.

Some of the church edifices are very fine specimens of ecclesiastical architecture, notably Christ Church (Episcopal), which is built in Gothic Cathedral form, and contains a handsome nave, adorned with stained glass windows. The Catholic Cathedral has a façade of polished freestone, with a Doric portico, and lofty spire, with a fine chime of bells. The First Presbyterian is a fine Gothic edifice, with a particularly graceful spire, and the Jewish Temple is one of the finest places of worship in the city.

The public-school system of St. Louis is excellent; the numerous school buildings are commodious, well-ventilated, and many of them really handsome in appearance. The Washington University is the seat of the higher education of the city, and connected with it, are the Mary

Institute for the instruction of females, the Polytechnic School, and the St. Louis Law School. The students number 700, and there are 60 teachers connected with the different departments. The St. Louis University, a Jesuit institution, is the oldest seat of learning in the city; it has a valuable museum, and a library of 17,000 volumes, containing some rare specimens of early printing. There are several good public libraries, the principal of which, called the Mercantile Library, contains a fine reading-room, with 45,000 volumes, and all the periodicals and magazines of the day, and collections of paintings, coins, and statuary.

The pride of St. Louis, however, is the noble bridge over the Mississippi, which at this point is rather narrow, and the current consequently very rapid. This bridge, justly regarded as one of the greatest feats of American engineering, is constructed in three cast-steel spans, (two of which are each 500 feet wide, the centre one being 520 feet), supported upon four granite piers, sunk over 100 feet through the sandy bed of the river, until they rest upon the solid rock. These spans or arches are sixty feet above the water-level, and therefore do not impede the navigation of the river; and the bridge itself consists of two roadways, an upper one for carriages and pedestrians, and a lower one for railway trains. The lower roadway enters a tunnel from the bridge, nearly 5,000 feet in length, which extends under a great portion of the city.

The cost of construction of bridge and tunnel was £2,000,000.

There are several very fine parks in St. Louis; Lafayette Park, the principal, is a fine piece of ground only thirty acres in extent, but being so well laid out, and intended for pedestrians only, it seems of much greater extent than it really is. Tower Grove Park, and Shaw's Garden, are also pleasant recreation grounds.

Bellefontaine Cemetery, the most beautiful in the Western States, embraces 350 acres, and is tastefully laid out, with beautiful trees and shrubberies, and contains many fine monuments.

As a manufacturing city, St. Louis ranks directly after New York and Philadelphia; its manufactured products amounting to an estimated annual value of £40,000,000. As the natural *entrepôt* of the whole valley of the Mississippi, however, it is the centre of the immense grain, live stock, and provision trades; and is the great distributing market for the cotton, lead, tobacco, wool, and hides, produced in that great district. It is also the greatest flour market in the country, and a great seat of the pork-packing industry.

Before proceeding South, I may say that Western Americans gave me the impression of being more active and energetic than their fellow-countrymen in the Northern and Eastern States; of being more broad-minded, less inflated with their own importance, more observant of passing events, and tolerant of difference of opinion. They do not appear to have the formality and self-conceit of the Yankee, nor the lassitude of the Southerner; and whilst attentive to their own affairs, yet seem very observant of those of the Northern, Eastern, and Southern States, and also to take an interest in foreign politics. They are certainly more national in their ideas than the New Englander, who thinks his own State alone worthy of notice; than the Eastern men, who think their ideas should alone dominate the country; and than the Southerners, whose attention seems to be solely devoted to recovering the position they occupied before the civil war.

It is in consequence of these characteristics of Western men, that careful observers conclude, that St. Louis will become the metropolis of the whole country; or that the Western States will in time be erected into an independent nation, with St. Louis as the capital. This latter seems the more probable, as the Western States already contain a population of eleven and a half millions; and when the land is all opened up and settled, their interests will undoubtedly be better served, by having a central government of their own at St. Louis, than by the existing one at Washington, which has to legislate for so many conflicting interests.

CHAPTER XVII.

DOWN THE MISSISSIPPI TO NEW ORLEANS.

THE Mississippi River—The " *Great Republic* "—Cairo—Columbus
—Hickman — Memphis — Helena — Napoleon — Vicksburg—
Natchez — Baton Rouge— River Scenery—New Orleans—
Position—History—Streets and Squares—Public Buildings—
Churches—Public School System—French Market—Cemeteries
.—Levee—Commerce.

HAVING decided to proceed to New Orleans by steamer
down the Mississippi, I took my passage by the *Great
Republic*, which was advertised to sail on the following
day.

Before describing the trip, a few words about the river
itself may not be out of place. The Mississippi, which
means "the Great River," literally "the Father of
Waters," rises in the highlands of Minnesota, in a cluster
of small lakes, near the sources of the Red River of the
North and the rivers that flow into Lake Superior. Its
sources are 1,680 feet above the Gulf of Mexico, into
which it enters. Its general course is southerly with
numerous windings and it has a length of 2,986 miles to
its mouth, from which to the source of the Missouri is
4,506 miles. The Mississippi and its tributaries drain
an area of 1,226,600 square miles. It is navigable to
the Falls of the St. Anthony a distance of 2,200 miles, or,

reckoning the Missouri with it, boats can proceed from its mouth a distance of 3,500 miles. It has 1,500 navigable tributaries, the principal of which are the Red River, 340 miles long from its mouth ; the Yazoo 534 miles; the Arkansas 700 miles ; the Ohio 1,053 miles ; and the Missouri 1,253 miles. The Mississippi averages for its whole course a width of 3,000 feet, and is from 75 to 120 feet deep. There is no apparent increase from the largest branches, and it is estimated that 40 per cent. of the flood waters are lost in the great marshes. Thousands of acres of land on the banks, are annually carried away by the current. The Mississippi forms a portion of the boundaries of ten States, having the southern part of Minnesota, Iowa, Missouri, Arkansas, and most of Louisiana on the west bank ; and Wisconsin, Illinois, Kentucky, Tennessee and Mississippi on the east. The chief towns situated on its banks are New Orleans, Natchez, Vicksburg, Memphis, St. Louis, Quincy, Keokuk, Galena and St. Paul.

And now a few words about the *Great Republic* which is the largest of the immense three-decked Mississippi boats. The saloon of the *Great Republic*, which is 260 feet long, is painted white and gold picked out with blue and is remarkably pretty. It has a double row of pillars with fretted arches forming three aisles, of which the side ones abut on the state-rooms. Round the saloon are covered galleries ; above is a tier of small apartments in which the officers and employés of the boat sleep, and above that, in the centre, the tower from which the vessel is steered. Over the paddle-boxes are a bar-room and a barber's-shop. The lower part of the vessel resembles a series of immense barns : here are the enormous engines, furnaces, and stores of coal and wood, piles of cargo, horses, mules, and other animals. In one corner is a carpenter's shop ; in another a blacksmith's forge. On deck towards the bows, is hung a fine deep-toned bell, which would put to shame many of the church bells one hears. Meals are served in the saloon at tables that accommodate about ten persons each.

Having noticed these various features of what is destined for some few days, to be my home, I proceed on deck to have a look at the country through which we are passing. The scenery is rather pretty ; low wooded hills from time to time approach the river on either side, and there are frequent signs of cultivation and habitation. At times there are pretty limestone bluffs, hollowed out in places into caves and arches, evidently by the action of the water at some remote epoch, when its bed was at a far higher level than it is at present. For thirty miles below St. Louis, the Iron-Mountain Railway runs along the river bank. The ore at this place is very rich, and almost pure.

We soon approached the city of Cairo, which is situated on the Ohio, just above its junction with the Mississippi. Cairo is built on a bank of slimy mud.

As the steamer approached the desolate embankment, which seemed the only barrier between the low land on which the town is built and the waters of the great river rising above it, it certainly was difficult to imagine that sane men, even though they be speculators, could have fixed upon such a spot, on which to place the site of a city—an emporium of trade and commerce. The town itself is a collection of brick houses and wooden shanties, and the streets are rendered almost impassable by mud. A more desolate-looking place cannot be conceived. Surely Dickens must have had Cairo in mind, when he described the flourishing town of Eden.

An hour and a half's journey from Cairo brought us to Columbus, which is situated on an elevated spur of land projecting into the water. The river here is very wide, in fact it did not appear to me to be wider at Vicksburg or Baton Rouge, which are not far from its mouth. On the hills behind Columbus may still be seen the dismantled ruins of strong earth-works, thrown up during the civil war to protect the town. A large island here impedes the stream, which runs swiftly under the bluffs, large portions of which become undermined and fall into the river.

A couple of hours after leaving Columbus we stopped at the desolate looking village of Hickman, which is on the "Ole Kentucky shore," at this point a very slimy one. The scenery of the river, if scenery it can be called, was now dreary in the extreme and continued so for the whole of the distance to New Orleans. Surely the Mississippi must be the most uninteresting river in the world, in spite of the boastings of oratorical patriots. Not a particle of romance can possibly attach to the immense forests of poor timber, or the dismal swamps which alternate with them.

The next day we reached Memphis, in the State of Tennessee, 420 miles below St. Louis. This flourishing new city stands on a yellow bluff, thirty feet above the highest floods, and is already a place of much importance. It extends for several miles along the high banks of the river, though it does not run far back. The streets are at right angles to the principal thoroughfares, which are parallel to the river. In the centre of the town is a green' square planted with trees, which seems a place of great resort by the citizens. The lofty stores and warehouses, the rows of shops on the broad street along the river, and the number and size of the public and private edifices, attest the results of the development of commerce created in a great measure by the Mississippi. Memphis is the outlet of a large cotton district, and exports 400,000 bales annually. It has fine public buildings and hotels, a theatre, eighteen churches, two medical colleges, five daily and three weekly newspapers, besides numerous banks and insurance offices. It is connected by railway with New Orleans, Charleston, Louisville, and Little Rock; and possesses foundries and manufactories of boilers and machinery. Its population is estimated at something over 30,000. During the War of Secession it fell into the hands of the Federal forces after the fall of Island No. 10, in 1862, and was the base of military operations for the capture of Vicksburg. Memphis is a wonderful place, and impressed me with the idea of progress more than any other place in the States. I was perplexed and amused by the mixture of whites, negroes, and of the semi-savage, degraded by his contact with the white man; by the contrast between the gigantic steamer and the "dug-out" of the black man, which are to be seen in close proximity on the river; by the roll of heavily-laden drays and the rattle of cars in the streets, and at all the phenomena of active commercial life, being included in the same scope of vision that takes in, at the other side of the Mississippi, lands scarcely yet settled, and some that remain in the same state as they were centuries ago.

Human life is still held cheap on the Mississippi, and "differences" still frequently occur, which end in bloodshed.

The next place of any importance we arrived at was Helena, a small town in the State of Arkansas. Helena stands in the mud, at the foot of some low hills, and was the scene of a severe engagement during the war. For a long time we continued our dreary way, until we arrived at Napoleon, a wretched-looking place, consisting of a collection of wooden houses, situated on a spit of muddy land, near the mouth of the Arkansas river.

The next day we reached Vicksburg, which is about 400 miles distant from New Orleans and 120 from Natchez, and which stands on a high bluff of yellow clay on the left bank of the river. Seen from the river with the remains of its great earth-works, and with the Court House and the spire of the Roman Catholic Church on the highest points, Vicksburg has a somewhat imposing and even picturesque appearance.

Vicksburg is the largest town, though not the capital of the State of Mississippi, and its exports of cotton before the war amounted to 100,000 bales per annum. It was strongly fortified in 1862 and provided with a numerous garrison. In January 1863 it was attacked by the Federal naval force from Memphis and New Orleans, but without success. In April 1863 a naval attack was combined with the land forces under General Grant, who defeated General Pemberton near Jackson, cut off supplies and reinforcements for the garrison, and with a close siege and continual assaults compelled a surrender on July 4th 1863, with 30,000 prisoners, 200 cannon, and 70,000 stand of arms. From the natural strength and importance of its position Vicksburg has often been called the "Quebec of the Mississippi," but the town itself is miserable, the streets for the most part being unpaved, and the buildings irregular, and generally constructed of wood or brick. From the terrace of the Court House, the view of the river, and of the vast tracts of forest, extending as far as the eye can reach, is, however, very fine. A little above the town, situated on a hill that slopes gently down to the river, is a cemetery in which repose the bodies of more than 30,000 Federal soldiers, an awful memorial of that terrible fratricidal war, called by some Americans "a little family quarrel." Vicksburg is a place of much commercial importance and contains about 10,000 inhabitants.

After leaving Vicksburg there was nothing to break the monotony of the next 100 miles, until we approached Natchez, when the banks became steeper, and the scenery a little more interesting. Natchez, distant from New Orleans 280 miles, is situated on a bluff 150 feet high, which here forms the river-bank. A portion of the town is called Natchez-under-the-Hill, and was formerly the resort of the river gamblers, pirates and other desperate characters. The city has ten churches, a Court House, Jail, the United States Marine Hospital, and possesses two daily papers. It is the shipping port of a large and fertile cotton district and has steam communication with the whole Mississippi valley. Natchez, which derives its name from a noted tribe of Indians, was settled by the French in 1716, and destroyed by Indians in 1729, but was subsequently re-built. Its population is estimated at 20,000.

. Shortly after leaving Natchez we passed the mouth of the Great Red River on the right, which came rolling out from amidst forests looking nearly as broad as the Mississippi itself; and yet the latter after its junction with it did not seem to gain at all in width.   The settlements on the river banks now became more numerous, and we stopped at several small places, one of which, Port Hudson, was prettily situated on a bluff of loamy clay and was very refreshing to the eye after the low swampy forests and flat plantation lands which border the greater portion of the lower Mississippi.

The next stopping place was Baton Rouge, a small town, formerly the political capital of the State of Louisiana.   Like Vicksburg and Natchez, it stands on rising ground, and is about eighty miles to the north-west of New Orleans.   It is a dull and sleepy place.   In the centre of the town stands the Capitol, a big castellated building, which was gutted by fire during the war ; since which time the Legislature of the State has held its sittings at New Orleans, and Baton Rouge has lost its pride of place, as capital of the State.   As far back as 1838 it was the seat of a college.   Besides the Louisiana Penitentiary, Baton Rouge contains an Asylum where all the deaf and dumb of the State from ten to thirty years of age, and all the blind between the ages of eight and twenty-five, are entitled to be educated and maintained at the public charge.   Baton Rouge contains about

18

10,000 inhabitants and on the opposite side of the river another town of the same name contains nearly an equal population.

From Baton Rouge to New Orleans the banks are flat and uninteresting and the country seemed to consist, for the most part, of low swampy land.

Take it altogether, the journey from St. Louis to the mouth of the Mississippi is one of singularly little interest. The desolation is oppressive. For hundreds of miles, dense forests of poor, weedy-looking trees alternate with undrained swamps. Towns are rare, and vast tracts of land intervene between them. The villages and detached shanties stand on unhealthy clearings, and the rotting timbers which support them are plastered with advertisements of specifics against chills, agues, and fevers. There are hundreds and hundreds of square miles of rich marsh and forest land waiting to be drained and so made to minister to the wants of a thriving population, and to the enrichment of the country at large. It seems almost incredible that instead of fostering this important object, the United States Government has recently thrown away millions in purchasing a wretched country like Alaska, from mere lust of possessing more territory.

New Orleans the political and commercial metropolis of the State of Louisiana, is situated on both sides of the Mississippi, but principally on the left, about 100 miles above its mouth. Though large, it is anything but a fine city, being built on the alluvial banks of the river, on ground lower than the high-water level, and only protected from inundations by a levee or embankment of earth, four feet high and fifteen feet wide, that extends for a great distance on both sides of the river, and forms a pleasant promenade in the winter months. The water that percolates through this embankment and the natural drainage is conducted by open gutters, which run through the streets, into a swamp that lies between the city and Lake Pontchartrain, three miles distant. There is always the danger of the Mississippi making a breach in the embankment and pouring its waters into the city; besides which, Lake Pontchartrain has a nasty habit of backing up and inundating it, after the prevalence of certain winds. Thus New Orleans is unpleasantly situated between two waters; and the soil is so full of moisture that no excavations can be made. The largest buildings have no cellars below the surface; and in the cemeteries there are no graves, the dead being placed in tombs above ground.

The older part of New Orleans is built within a great bend of the river, from which circumstance it derives its name of the "Crescent City." It has however long ago overstepped its original limits, and now extends for a distance of about twelve miles along the river bank, presenting an outline somewhat like the letter S.

New Orleans was settled by the French in 1718, but was abandoned in consequence of floods and sickness. Another, and more successful attempt at settlement was made in 1723, and the colony was held by the French until 1729; then by the Spaniards till 1801, and by the French again until 1803; when it was ceded with the Province of Louisiana to the United States. In 1860, Louisiana having seceded from the Union, New Orleans became an important centre of commercial and military operations, and was closely blockaded by a Federal fleet. An expedition of gunboats under Admiral Farragut forced the defences at the mouth of the river on April 24th, 1862; when the city was forced to surrender, and was occupied by General Butler as military Governor.

The streets of New Orleans are very wide and handsome in appearance, though only the principal of them are paved. Those parallel with the river extend in an unbroken line, for a distance of about twelve miles; those at right angles to them, that run from the river to the lake, are also very regular. The streets that are not paved, are simply quagmires: in winter they are not practicable at all, and even in summer the dust makes them almost impassable. The open gutters form a bad feature of the streets of New Orleans; these have very steep sides, and are crossed at street corners by small bridges, consisting of single stones, and allowing two persons only to cross at a time.

Canal street is the main business thoroughfare, and promenade ; and may be said to divide the city into two pretty equal parts. It is nearly 200 feet wide, and has a grass-plot twenty-five feet wide, and bordered with two rows of trees in the centre, extending its whole length. Claiborne, Rampart, St. Charles, and Esplanade streets, are embellished in the same manner. In Canal street is a colossal bronze statue of Henry Clay.

Jackson Square is the favourite place of resort ; it contains beautiful trees and shubbery, and shell-strewn walks ; in the centre stands an equestrian Statue of General Jackson. When the Federals occupied New Orleans, they, with execrably bad taste, cut twice upon the granite pedestal of this statue, the motto " The Union must, and shall be preserved," making it appear as if General Jackson had enunciated that sentiment. Overlooking this square, is the French Cathedral of St. Louis, built in the old French style, and two Court Houses in the Tusco-Doric ; which have a very picturesque appearance.

Lafayette Square is also a handsome enclosure, and contains a fine marble statue of Franklin by Hiram Powers. The City Hall, Oddfellows' Hall and a fine Presbyterian Church, all front this square.

New Orleans is not remarkable for its architecture, but it possesses a few very fine buildings, the principal of which is the Custom House and Post Office. This fine edifice is built of a dark granite, and is after the Capitol at Washington, the largest building in the States ; it has some fine columns with heavy Egyptian capitals, and the Long Room is a very handsome hall. The City Hall is certainly the finest edifice in the city ; it is built of white marble in the Ionic style, with a wide and high flight of steps leading to an elegant portico, supported by eight columns. The State and City Libraries occupy rooms in this building.

The churches of New Orleans are numerous and handsome. The most famous is the Roman Catholic Cathedral of St. Louis, which has an imposing façade, surmounted by a lofty blue-slated steeple and flanked by two towers, each capped by a smaller blue-slated spire. The paintings on the ceiling of this building are by Canova and Rissi. The finest Episcopal Church is St. Peter's, which is a handsome specimen of Gothic architecture, and has a very rich interior. The Presbyterian Church is a fine structure in Greco-Doric style, and is much admired for its fine steeple. The Temple Sinai, the principal Jewish place of worship, has a light and elegant appearance ; it is built of parti-coloured bricks, and has a handsome portico flanked by two towers, surmounted by tinted cupolas. Its Gothic windows are filled with beautifully stained glass, and the interior is remarkably rich and beautiful.

There are in New Orleans eighty public schools, and

numerous private ones, mostly Roman Catholic, which provide for the instruction and moral training of the rising generation. Many of these are high-class educational institutions, the principal being the University of Louisiana, which has only the two departments of law and medicine, but these are of a high order, and very well attended. The medical college, contains a fine anatomical museum and other collections. Straight University is exclusively for coloured students, and gives instruction of good grammar-school grade. The public school system of New Orleans is secular and free. The schools are divided into " High schools," " Grammar schools " and " Primary schools" and are governed by a board of directors chosen by the City Council, who levy a special tax for educational purposes. Opposition to the system has only proceeded from the Roman Catholic body, who are, of course, very strong in New Orleans. They have not, however, as elsewhere, contented themselves with simply denouncing the schools as " Godless," but they have erected magnificent schools themselves, in which children receive a good education for one dollar a month, or, gratis, if they cannot pay that sum.

The principal charitable institutions of New Orleans are the Charity Hospital and the Hôtel Dieu. The former is one of the finest buildings in the city, and one of the noblest institutions in the country ; it was founded in 1784, has stood on its present site since 1832, and has accommodation for 500 patients. The Hôtel Dieu is a fine hospital, established by the Sisters of Charity and supported entirely by receipts from patients, although many are admitted free of charge.

A visit to the French market, which comprises several buildings on the levee, is most interesting. The market people commence to assemble at daybreak, and it appears as if all nations and tongues have their representatives in the motley and ever-moving crowd. The noise, however, is far from being unpleasant to a visitor's ear. The prevailing language is French, and is heard in every dialect and patois, from the fluent and musical accents of the polished Creole, to the childish jargon of the negroes. The articles exposed for sale are infinite in their variety, but the fruits and flowers are especially attractive. The former embrace all the products of both the temperate and torrid zones; and the rich colours of the flowers are wonderful to behold.

The cemeteries of New Orleans are noteworthy for the peculiar mode of interment in them. From the nature of the soil, which is semi-fluid at a depth of two or three feet below the surface, all the tombs are above ground. Some of these are very costly and beautiful structures of marble and stone, but the great majority only consist of cells, placed one above the other, generally to the height of seven or eight feet. Each cell is only large enough to receive the coffin, and is hermetically bricked up at its narrow entrance as soon as the funeral rites are over. In most instances a marble tablet, appropriately inscribed, is placed over the brickwork, by which the vault—or oven, as it is locally termed—is closed.

The levee is one of the most characteristic sights of New Orleans, and for extent and activity it has no parallel on the continent. Here a thousand river steamers and flat boats may be seen at one time ; whilst its wharves are lined with hundreds of sailing and steam ships from all parts of the world. New Orleans commands 20,000 miles of steamboat navigation, and is the natural *entrepôt* of one of the richest regions on the continent. In the value of its exports, it ranks after New York ; it is the principal cotton mart of the world, and besides cotton, it ships sugar, tobacco, flour, pork, &c., to the total value of £20,000,000. Its imports, which consist principally of coffee, iron, salt, drapery, and spirits, amount to £3,000,000. Its manufactures are unimportant, and its population is estimated at 210,000 inhabitants.

CHAPTER XVIII.

MOBILE, SAVANNAH, CHARLESTON, RICHMOND.

LAKE Pontchartrain—Description of Mobile—Harbour—Indians—
Vicinity — Montgomery — Atlanta — Macon — Description of
Savannah—Pulaski Monument—Bonaventure—Rail to Charles-
ton—Position—Harbour—Description of the City—Public
Buildings—Ruined Plantations—Columbia—Wilmington—The
City of Richmond—Capitol—St. John's Church—Statue of
Washington—Condition of the South.

LEAVING New Orleans by the Pontchartrain railway, I
found myself within an hour on the steamer proceeding
down Lake Pontchartrain, *en route* for Mobile. This
lake is distant about five miles from New Orleans ; its
muddy waters teem with fine fish, and are covered with
game of all kinds ; it is forty miles long, twenty-five
miles wide, and from sixteen to twenty feet deep. The
shores are covered with dense forests of fine timber,
chiefly pine and cypress, and abound with deer. The
trip was somewhat monotonous, until we entered the
channel called the Rigolettes, and passed Fort Pike,
when the scenery became a little more diversified. In
the course of a few hours we were in the Mississippi
Sound, catching occasional glimpses of the open Gulf of
Mexico ; and soon entered the Bay of Mobile. Here,
below the bar of the Alabama River, at a distance of

twenty-five miles from the city, vessels drawing over ten feet of water are compelled to lie, their cargoes being conveyed to and from the city in small steamers; but improvements are now being made, that will enable vessels of thirteen feet draught to get up to the wharves.

Mobile, known also as the "Gulf City," is the largest town and the only seaport in the State of Alabama; and is situated on the Mobile River, thirty miles distant from the Gulf of Mexico, in the midst of a sandy plain, bounded at the distance of a few miles by high hills. Though regularly laid out, with well-paved and delightfully shaded streets, I should say it is one of the dirtiest and most dismal towns in America. Government street however, the principal promenade, is a fine thoroughfare shaded by grand oaks, and containing many handsome private residences with beautiful gardens. There is too a public square, its walks well shaded by fine trees, that is also a nice feature of the town, and tends to modify its general dreariness. The only building in Mobile that calls for special mention is the Custom House, which also contains the Post Office; this is a handsome edifice built of granite.

The City now contains a population of 40,000, and its principal business consists in the shipment of cotton, of which staple 350,000 bales are annually exported. There are also a few foundries and machine shops and other industries in the town.

The harbour is strongly defended, and during the late Civil War the fortifications were attacked by the Federal fleet under Admiral Farragut, who ran the gauntlet of the batteries, destroyed the Confederate fleet including the ram *Tennessee*, but did not succeed in capturing the city itself, which only fell after the surrender of General Lee, and was one of the last Southern towns occupied by the Federal troops during the war.

In the outskirts of Mobile there is a small settlement of Choctaw Indians who inhabit huts, open on one side, constructed of bark and covered with deerskins, and so low that they can only be occupied in a sitting posture. These Indians who live chiefly by cutting wood, which is sold in the city by the squaws, though some of them hunt, are a fine race, being tall and well-made, with bright black eyes and light coppery complexions. They seem extremely poor and miserable, and will no doubt in the course of a few years entirely die out, unless something is done to improve their condition.

Spring Hill is a pleasant suburb some six miles distant from the city, where there are a number of pretty villa residences embowered among the woods. In close proximity is a fine forest of pines, oaks, chestnuts and giant magnolias, which latter are quite equal in size to the other large forest trees, and, with their brown trunks, masses of green foliage, and lovely white flowers, are certainly pre-eminent in beauty.

Much lawlessness seems to prevail, and crime is of frequent occurrence, owing in a great measure to the still unsettled state of the South, and the political ascendancy of the negroes; but on this subject I will write at greater length later on.

The heat was so intense that I did not make a long stay in Mobile, but taking the Mobile and Montgomery Railway, soon left for the latter place. Montgomery is the political capital of Alabama, and in size and importance the second city in that State. It is situated on the Alabama River and contains a population of some 13,000 inhabitants. A prominent feature in it, and one visible for some distance around, is the Senate House situated on an eminence called Capitol Hill; it is a fine building with a large dome. Montgomery was the first capital of the Southern Confederacy, but was afterwards supplanted by Richmond.

I did not remain long in this town, but availed myself of a train that started for Atlanta a couple of hours after my arrival. The route lay through the most thickly settled districts in the States of Alabama and Georgia, past many pretty little towns, and though the scenery was not strikingly picturesque, it yet possessed the Southern characteristics of luxuriant vegetation to relieve it from monotony. Atlanta is the capital of Georgia, and after Savannah the largest town in that State. Its importance arises from its position as the centre of several railway

systems that here converge, and it has many of the
features of those Northern towns that owe their rise to
the same cause. It is however picturesquely situated,
and possesses a few good buildings, the principal of
which are the State House, the City Hall, and the Union
Passenger Depôt.

Being desirous of hastening on to Savannah, where I
intended to stay for a few days, I did not wait at Atlanta,
but at once proceeded by rail to Macon, which place I
reached after a journey of five hours, through country
similar in appearance to that between Montgomery and
Atlanta. Macon is a remarkably pretty place on the
Ocmulgee River; it is regularly laid out, and is quite
embowered in trees and shrubbery. It is one of
the most prosperous towns in Georgia; contains a
population of some 10,000 inhabitants, and possesses
important industries in its iron foundries, machine shops,
and flour-mills. After leaving Macon, a journey of some
ten hours' duration brought me hot and dusty to
Savannah, where I found good accommodation at the
Screven House and enjoyed the luxury of a bath, which
by the way always costs two shillings throughout the
States.

Savannah is without doubt the handsomest city in
the South and perhaps in the whole of the States; for
it is embowered in the luxuriant foliage of orange-trees,
bananas, magnolias, stately palmettos, flowering oleanders,

pomegranates, myrtles, bay and laurel trees. The private residences have mostly beautiful gardens, which are in constant bloom, and the city itself is laid out very regularly, the streets being wide, well-shaded, and crossing one another at right angles. A charming feature of Savannah too is the number of squares or greens at the intersection of the principal streets. These squares, twenty-four in number, vary in extent from two to three acres, and being situated equi-distant from one another, laid out in walks, and planted with evergreens and ornamental trees, assist materially in giving to the city, that appearance of tropical luxuriance of vegetation, that constitutes its great beauty.

The position of Savannah, from a commercial point of view is very good. It is situated on the Savannah River, about eighteen miles above the point where it empties itself into the Atlantic; and it has become the second port in the States for the shipment of cotton. Its river-front is in the form of a crescent, extending a distance of about three miles, and all the large warehouses are here built on a narrow strip of land, that intervenes between the wharves on the river-front and the base of a steep bluff; thus their uppermost windows in the rear overlook a sandy plain on the top of this bluff, planted with rows of trees, which, under the name of "the Bay," is the great commercial mart of the city.

Most of the public buildings, which however are not

particularly noteworthy, are erected on land fronting the
different squares in the city.   They are consequently very
conspicuous, the most prominent being the Custom
House and Post Office, which is a fine granite edifice.

In one of these squares is a fine monument erected to
the memory of Count Pulaski, who fell in 1779, in the
attempt of the combined French and American forces to
recapture the city, then in the hands of the British, who
had taken it by assault in the preceding year.   This
chaste monument, erected on the spot where Pulaski fell,
consists of a marble shaft fifty-five feet high, surmounted
by a figure of Liberty holding the national flag.   In
another of these squares there is a Doric obelisk
commemorating General Greene and Count Pulaski
jointly.

At a distance of four miles from the city, on the Warsaw
River a branch of the Savannah, is the beautiful
Bonaventure Cemetery, which, at one time the private
estate of an English family, is now used as a resting-place
for the dead ; Nature having seemingly intended it for
that purpose.   While still in private hands, it had been
laid out in avenues of live-oak trees, and these have in
course of time assumed the proportions of forest giants,
and stand like colossal columns on either side ; whilst
their ever-green foliage, interlaced high overhead, excludes
the light, and gives to the natural aisles thus formed, a
sombre and solemn aspect, quite in harmony with the

purpose to which they are put. Nowhere have I seen so appropriate a home for the dead, as in these darkened leafy glades, where the very moss and wild vine that hang pendent from the green canopy overhead, seem to mourn for those who lie buried beneath their shade.

After spending a couple of days very pleasantly, in spite of the great heat at Savannah, I proceeded by the Savannah and Charleston Railway to the latter place. The road thither runs within a few miles of the sea-shore, but we never caught sight of the ocean, and for a great distance passed over swamps, and across numerous muddy streams, the rails being laid on piles. The scenery though not picturesque, was rich in grand vegetation, for we passed many dense forests of pines, cypresses, bay and laurel trees; while huge oaks of enormous size, that must for centuries have withstood the ravages of time, lined the road on either side; in parts forming magnificent avenues, within the shade of which, innumerable flowers of all colours and hues made beautifully variegated floral carpets.

Charleston is situated on a narrow spit of land formed by the junction of the Ashley and Cooper Rivers, at the point where they enter the sea; it thus possesses three water frontages. It is now the principal city in South Carolina in size and commercial importance, and contains over 50,000 inhabitants. It is, for America, an old city, having been settled by an English colony in 1679. It

19

played an important part in the Revolution, which obtained for the States their independence, and also in the first stages of Southern secession, the cause of the late civil war.

The Harbour is an estuary about seven miles in length, extending to the Atlantic; it is almost land-locked, the entrance being only a mile in width. It is protected by strong batteries, the principal of which are Fort Moultrie, Fort Sumter, and Castle Pinckey; the former being situated on Sullivan's Island, at the entrance to the harbour on the right. Fort Sumter is now in ruins, and forms a picturesque feature of the harbour view; it is built on a shoal on the left side of the entrance, and commands the channel by which vessels enter. Castle Pinckey, covering the crest of a mud bank, is immediately in front of the town, being distant from it about a mile, and directly facing the entrance. It was here that open hostilities first commenced in the late war, when Fort Sumter was bombarded by the Confederates, who compelled the Federal garrison to surrender; and for a long time afterwards these fortifications were one of the chief points of Federal attack.

Charleston is built on low flat land; extends a distance of three miles from north to south, and is laid out with some degree of regularity. There is no uniformity in the buildings, but the absence of regularity in this respect is more than compensated for, by the greater diversity;

and as the houses are mostly detached, quaint in appearance, and surrounded by gardens containing the grand oaks, magnolias, and that luxuriant vegetation peculiar to the South ; the town has a most picturesque appearance. Charleston suffered very much during the civil war, but since then, such progress has been made in the work of rebuilding, that it now shows but few traces of the great damage it sustained.

King street is the principal thoroughfare and contains the best shops. In Meeting street are the warehouses in which the wholesale trade of the city is conducted, and the Banks and Insurance Offices are located in Broad street. The prettiest feature of Charleston, is a fine promenade called the Battery, situated at the water's edge, and surrounded by the best private residences. From it a magnificent view of the harbour is obtained.

The country to the north of the city is very beautiful, and the drives in that direction along the banks of the Ashley and Cooper Rivers are most enjoyable, passing as they do, through rich tropical shrubbery.

Charleston possesses numerous public buildings ; they are however with a few exceptions not remarkable for beauty of design. The new Custom House which is not yet completed, is a handsome edifice of white marble, in the Corinthian style, with a very graceful portico. The City Hall too is a fine building, approached by a double flight of marble steps. The Orphan House, standing in

the middle of extensive grounds which contains a statue
of William Pitt, erected during the revolutionary times, is
one of the most prominent buildings in the town, and
one of the best institutions of the kind in the States.   It
has the honor of having produced men, who have risen
to distinction and attained the highest positions.

There are two churches in Charleston that are noteworthy on account of their comparative antiquity.   These
are St. Michael's and St. Philip's ; the former of which
was erected in 1752, from the designs of a pupil of
Sir   Christopher   Wren,   and   possesses   a ʼfine   spire,
which forms a prominent landmark even far out at sea.
St. Philip's is not quite so old, and the greatest interest
attaches to its graveyard, where lie buried the bodies of
the most illustrious of South Carolina's sons.

It is sad to see the devastation caused by the war on
some of the plantations in the neighbourhood, notably
on James Island ; but nowhere does this feeling of
melancholy so forcibly oppress one, as during a visit to
Middleton Place, formerly one of the most beautiful
plantations in South Carolina, and which still exhibits
in its luxuriant shrubbery and magnificent old oaks,
its lakes, and picturesque old tombs, traces of its
former glory, now, alas, wrecked and ruined by the
unsparing hand of war.

Leaving Charleston I started on my way to Richmond,
taking the cars of the Atlantic Coast line, and completed

the long journey of 570 miles in a day and night. As far as Columbia, distant from Charleston 130 miles, the scenery possessed some of the Southern characteristics of rich vegetation, though in its least beautiful form ; since the country is flat, and covered with extensive forests of pine trees. Columbia is the capital of South Carolina and is picturesquely situated on the bluffs of the Congaree, just below the lovely falls of that river. In 1865 during its occupation by General Sherman's forces it was the scene of a great conflagration, which destroyed the magnificent gardens and fine trees shading the streets, that constituted its most beautiful features. It is still a remarkably pretty place and one of some importance, containing a population of some 9,000 souls. There is in course of erection a Capitol or State House that will be when completed amongst the handsomest buildings in the States. The Asylum for the Insane is also a fine edifice.

Continuing the journey from Columbia ; the country through which the line passed became more monotonous, though presenting the same general characteristics of flat country, with numerous belts of pine trees. The stations on this portion of the road were few and far between, the principal being Sumter and Florence. After travelling about 110 miles we arrived at Wilmington. This is the largest city in North Carolina, and is situated on the Cape Fear River, about twenty miles from its mouth, which is defended by a strong battery called Fort Fisher.

From Wilmington the scenery continued of the same uninteresting character, and throughout the long stretch of road, over which we had to travel, before arriving at Petersburg, 160 miles distant, we only passed two stations of any importance, Goldsboro, a prosperous little town of 5,000 inhabitants, situated at the head of navigation on the Neuse River, and Weldon a thriving little place in North Carolina on the Roanoke River. Petersburg is noteworthy, from having been the scene of the last struggles of the Confederates in the late war, which resulted in its evacuation by General Lee, and led to the capture of Richmond by the Federals. It has prospered since the war, and the vestiges of the great battles that here took place are gradually being effaced, but the remains of the fortifications are still visible and form a mournful momento of that sanguinary fratricidal war. After passing Petersburg we crossed the James River on a fine bridge and soon arrived at Richmond.

Richmond the chief city and political metropolis of Virginia, and during the civil war the capital of the Southern Confederacy, is situated on the James River, about 100 miles by water from Chesapeake Bay. During the war, great importance was attached to its possession by both the Federals and Confederates, and the former often attempted its reduction. The obstinacy however, with which it was defended, may be seen in the remains of the strong line of earthworks thrown up around it, and it was only surrendered after General Lee evacuated Petersburg in 1865. To prevent the tobacco warehouses and public stores from falling into the hands of the Federals, they, together with the bridges over the James River, were destroyed by fire. In this manner a considerable portion of the city was burned; but directly after the cessation of hostilities, the work of rebuilding commenced, and Richmond now presents but few tokens of the conflagration, and is rapidly regaining its former prosperity.

The city, which is laid out with great regularity, is built on two eminences called Richmond Hill and Shocktoe Hill, separated by the Shocktoe Creek. Crowning the summit of the latter of these hills, and standing within a small park of some eight acres, is the State Capitol, a fine edifice with a portico supported on Ionic columns, the design for which is said to have been furnished by Thomas Jefferson, after that of the *Maison Carrée*, at Nismes, in France. Within this building is a life-size marble statue of Washington, erected by the Legislative Assembly of Virginia.

Richmond contains some good public buildings, the best and most prominent being the City Hall, and the Custom House and Post Office, both of which are handsome structures. There are also several buildings, to

which much interest attaches, from having been prominently connected with the late Civil War. Such are
the Brockenbrough House, formerly the residence of
Jefferson Davis when President of the Southern Confederacy, and the Libby and Castle Thunder Prisons, in
which so many Northern prisoners languished in con-
finement.

St. John's Church, too, which is a plain edifice, and
dates from ante-revolutionary times, is noted for its
associations with that period, for in it was held the Con-
vention to decide upon what course of action the Colony
of Virginia should take in the crisis that had arisen in
the relations between the thirteen Colonies and the
mother-country. On that occasion, it was Patrick Henry's
great speech that mainly contributed to the decision
being arrived at, to cast in its lot with that of the other
Colonies.

The Monumental Church, a handsome edifice with a
fine dome, is erected on the spot where formerly stood
the Richmond Theatre, which was consumed by fire
during a performance, when a great number of people
were killed. It is to commemorate that sad event, that
the Monumental Church was built.

Capitol Square, as the small park in which the Capitol
stands, is called, contains a fine equestrian statue of
Washington. This consists of a massive granite pedestal,
surrounded by bronze figures of Patrick Henry, Thos.

Jefferson, John Marshall, George Mason, Thomas Nelson and Andrew Lewis, surmounted by a colossal bronze figure of Washington on horseback. In this square there is also a life-size marble statue of Henry Clay, and as the little park is well laid out, it forms the great place of resort of the citizens.

The population of Richmond now amounts to over 60,000, and it possesses a large commerce, its principal articles of shipment being tobacco and flour. It is the centre of several industries giving employment to between four and five thousand men, but its chief support is the trade in tobacco.

The scenery in the vicinity is very beautiful and one of the loveliest spots is Hollywood Cemetery, in which, under the shade of its noble trees, repose the bodies of so many Confederate soldiers, to whose memory a monumental stone pyramid has been erected.

Three fine bridges span the James River, and connect the city with Spring Hill and a pretty little village called Manchester, which contains a couple of cotton mills.

At Richmond I embarked in one of the steamers of the Old Dominion Line and after a rapid and pleasant run, arrived at New York, having completed a round trip of some 4,500 miles since leaving it two months previously.

Before taking leave of the South I will briefly mention the causes that led to its present condition, and what I found that condition to be.

After the civil war was at an end, and the Federals began to reap the fruits of their victory, the confiscation of the estates of those who had taken part in or abetted the Secession movement, commenced. The ·result was, that a great portion of the property in the South changed hands, and families that had hitherto been living in affluence, were reduced to poverty. Plantations were ruined, towns in whole or part destroyed, and havoc and ruin, that always follow in the track of con-quering armies, were universal. Thus the South, that had strained every nerve and exhausted all its resources in the late struggle, from which it had emerged torn and bleeding, was still more reduced and humiliated, and left in a yet more prostrate condition, by the action of the Federal authorities, that is, of the dominant Republican party ; taken quite independently, and in defiance of the sanction of the laws. This was the time chosen for the emancipation of the slaves, and the Southerners had to learn the bitter lesson, that there was a still lower depth of humiliation for them, in having their ex-slaves suddenly converted, not into equals, but into masters and tyrants ; for at the time the slaves in the South were emancipated, the franchise was extended to them, long before it was conceded to the better educated free negroes of the North. At the time, too, that the suffrage was granted to the ex-slaves of the South, many of the whites, who had taken part in the war were deprived of their votes; and many, sick at heart at the existing state of affairs, migrated to the North : thus the negroes gained the political ascendancy, which they have hitherto maintained by force of numbers, and in spite of the disabilities having been removed from the whites.

The state of the South at present is, that the high-spirited whites, who made so gallant a struggle against such overwhelming odds, are now tyrannized over by their former slaves, and Northern agitators and demagogues called "carpet-baggers," who are foisted into power by the negroes, to whose ignorance they pander. Education and intelligence are thus prostrate before crass ignorance, and the bitter feeling existing between whites and blacks is exhibited in deeds of violence and bloodshed, which causes a large amount of lawlessness to prevail.

This hostile feeling between the two races is said to be fomented by the Republican party for the purpose of retaining the Negro vote. This may or may not be the case ; but Southerners certainly repudiate with scorn, the idea of there ever being a war of races ; and profess their willingness to live at peace with their black fellow-citizens ; if Northern agitators were prevented from inoculating them with the communistic theories employed to keep alive the agitation in favour of the Republican party.

The Negroes, who form the peasantry of the South, are lazy and generally steeped in poverty, which will go far towards decreasing their numbers, especially as it is confidently stated that since the war, the number of births has greatly decreased.

That the South will ever renew the struggle for independence seems very doubtful, and as the old election cry of the Republican party, raised for the purpose of keeping down the South, and suggestively called "the waving of the bloody shirt," appears to be losing its potency, it is to be hoped that the existing state of affairs will soon be amended.

CHAPTER XIX.

NATIONAL CHARACTERISTICS.

ENERGY—Silence—Exclusiveness—Extravagance in Language—
Extravagance in Dress—Low Tone of the Press—Absence of
Pauperism—Power of Assimilating Foreign Immigration—
Diffusion of Education—Sobriety—Speech—Tobacco-Chewing
—Notes on Religion.

WHAT first impressed me, after being a short time in
America, was the absence of marked distinctive traits in
the people, and their great resemblance in most ways to
Englishmen; for much as Northerners may differ from
Southerners, and they again from Western men; their
Englishness, if I may be allowed the expression, is still
very apparent. The cities too, especially in the Eastern
States, are essentially English in their appearance; and
a visitor to such towns as Philadelphia, Boston, and
Baltimore, might very easily imagine himself to be in
Manchester, Liverpool or Bristol. In Southern cities,
such as New Orleans, this is not so much the case, as
their English characteristics have been modified in
accordance with the climate. Western cities like Chicago have, from their newness, a more American appearance; that is, having been built in more recent times,
they have been laid out more regularly; in many cases

losing in picturesqueness what they have gained in uniformity.

One of the most prominent characteristics of Americans, and one that at once strikes a visitor, is their tremendous energy. No enterprise seems to be too great for them to undertake, and no obstacles in the way of success are allowed to daunt them. It is this energy that has re-built Chicago and Boston in so incredibly short a time ; that has formed great cities where but a few years ago, the axe of the pioneer was ringing amidst virgin forests ; that has completed an unrivalled railway system, extending through thinly populated districts, thus inducing settlement on the land ; and that has carried a line of rail from the Atlantic to the Pacific, a distance of more than 3,000 miles, over obstacles that might well have been deemed insurmountable.

In spite of what I had always heard to the contrary, the Americans appeared to me to be a silent people. In few instances, either in the public cars, in the railway carriages, on steamboats, or on other similar occasions when numbers of people were brought together, did I find that I was first addressed. If I even opened a conversation with my neighbour I usually received at first monosyllabic replies, until he learned that I was a stranger visiting the country, when he would become very communicative.

The better class are exclusive and essentially aristocratic in their tendencies; they keep aloof from actual participation in political life, in order to avoid contact with the class of people who unfortunately are in office in the States; and this in spite of their great stake in the country and consequent interest in the maintenance of order. There is a section of the people, generally belonging to the immigrant class, which is given to exhibiting its independence and equality, in season and out of season, and sometimes in a somewhat offensive manner. Social equality is very admirable, and I, in common with most, am very willing to concede it, as far as it can be carried out; but it is very objectionable to have it thrust upon one in an obnoxious manner, which is only a form of impoliteness, and tends to destroy those amenities of social life which are at once a necessity and ornament of civilized communities. On the other hand "flunkeyism" prevails to a great extent, the people generally having an undue admiration for wealthy men, and being almost servile in their adulation of members of Royalty and persons of distinction from the Old World. It was a common thing to see a crowd of several hundreds of people waiting to see the Emperor of Brazil enter his carriage; and when I was on board the *Inman* steamer leaving New York harbour, a salute of thirteen guns was fired from the batteries in honour of an Indian Major-General then on board. The newspapers too, chronicle the movements of the "upper

ten" with far more diffusiveness than do their proto-
types in England, and they generally contain a whole
column devoted to such fashionable news as :—the
lovely brunette Miss B. being about to proceed to
Saratoga for the season; as Senator C. having taken
a house for the season at Newport; as the Honourable
A. D. being about to take up his residence, together with
his charming wife, at his country-house on the Hudson,
where he intends entertaining a select circle of his
friends; and so on *ad nauseam.* In the Southern States
the pride of descent is very great, but is never exhibited
in an offensive manner. In the Northern and Western
States however, there prevails a great love for titular dis-
tinctions of all kinds, and it is common to give people
titles to which they have no claim. I was often addressed
to my astonishment as " Captain" and " Squire"; and
found that people generally were most punctilious in their
use of the " Sir," which they emphasize and use with too
much frequency. This national trait struck me as being
very much opposed to the Republican idea of equality,
and is held to be a strong argument in favour of the
opinion, that Imperialism may yet carry the day in the
States.

Americans, in speaking, use strong expressions to
describe ordinary events; everything in the country being
with them either the greatest and biggest in the world, or
so small as to be beneath criticism. They are most

extravagant in their praise of individuals—say of the leaders of their own party, and unmeasured in their abuse of members of the opposite one. A man, whose only crime may be belonging to a different political party, will have all his private affairs raked up, as if they were public property, and be spoken of in terms more suited to the description of a great criminal. To the supporters of his own party, a political leader will be a demigod—to his opponents a fiend.

The extravagance in dress is equally as great as the extravagance in language, and is not confined to the female portion of the community, but is common to both men and women. I would say that in no country in the world are the people so well dressed as in America, for even in a crowd collected for any purpose, political or otherwise, where the dregs of society, at other times hidden away in the back slums of the cities, appear on the surface, it would be difficult to find even one individual clad in such rags and tatters as are, unfortunately, so often seen in the large cities of the Old World. Amongst American women, the passion for dress is a species of monomania; they have certainly great taste in the choice and mode of wearing their attire, but it is equally certain, that in the majority of cases its excessive cost is not warranted by, or in conformity with their station and means. A great deal of the prevalent commercial immorality may be traced to this passion for dress, combined with the general high cost of living ; for heads of families who might otherwise make a fair income, and jog along very comfortably, in order to enable their wives and daughters to enter a little into society, in which case they have to " keep up an appearance," are compelled to toil from morning till night, and in numberless cases are induced to enter into speculations outside their legitimate businesses, for the purpose of increasing their incomes until they become sufficient for their greater requirements.. This, as can readily be seen, gives rise to much over-trading and consequent fraud. People have been pointed out to me at Saratoga and Long Branch who were known to have lived quietly for a couple of years, putting by por- tions of their income, and then to have launched out into fashionable life, where for a season or two they would be very prominent, and then having come to the end of their tether, would retire from the scene and settle down into their former quiet life.

It seemed to me that in the States everybody lives at high pressure, and that there is a constant craving for excitement. The newspapers recognize and pander to this general taste, for their articles are as a rule written in a highly-spiced and most sensational style and are pre- faced by exciting headings. A low tone pervades the press generally, and personalities are freely indulged in. The foreign news is insignificant, as if to the majority of readers it had no interest; and as regards articles written upon any event of interest occurring in Europe, the comments show such want of knowledge of the subject, and are couched in such self-gratulatory and patronizing language—as if the writer were rather amused at the vain efforts made to attain to the American standard of perfection—that they are quite beneath criticism. This is a great drawback to the pleasure derived from a visit to the States; for during his stay, a visitor is literally cut off from all that is taking place in the world; and on his return to Europe he will find that as regards the history of passing events, his absence has occasioned a perfect blank in his mind.

It would seem as if the vastness and diversity of their own great country were all-absorbing to Americans, and gave rise to this feeling of self-sufficiency; which, however, cannot be desirable so long as they are dependent upon Europe for so much of what embellishes their daily life. How the low tone and virulence of the press affects the political life of the country will be shown later on.

A pleasing characteristic of the States is the absence of pauperism, and this must be more apparent to English eyes than to those of Colonists, who are accustomed to the same feature in their own countries. In New York and some other of the large cities of the Eastern States, I was several times accosted by beggars; but this is certainly the exception, and not the rule, and I may safely say that throughout the Western and Pacific States I never saw a mendicant. No doubt, in the large cities there is a pauper population, however small; but it is not apparent, and only consists of those who prefer what Americans call a "loafing" life to the healthier and happier one of labour, that is open to all, and by which, those who are willing to work can obtain a good and respectable livelihood.

A surprising feature is the rapidity with which the foreign immigration is assimilated, and many conflicting elements fused into a distinct nationality. This process of assimilation is also seen, though in a lesser degree, in our own Colonies. It is often stated, however, by well informed Americans, that it does not continue in the same ratio that it formerly did; and that in consequence the large German and Irish immigration is not so readily absorbed, but preserves more or less its distinctive characteristics. If such be the case, and I saw no reason to doubt it, it seems fortunate for the future of the Great Republic that these two foreign elements are antagonistic to, and thus tend to neutralize one another.

The attention that is paid to education is most praise-worthy, although the standard is not high; and it struck me as being too utilitarian, deficient in refining influences, and tending to inculcate a certain narrow-mindedness. It is, however, generally diffused throughout the country; and though the better class does not attain the high standard of the corresponding English middle class, the lower classes are undoubtedly better educated and far in advance of our own.

In the above remarks I have referred to the education obtained in colleges and schools, but I must not omit the education by means of the numerous galleries of art, libraries, and scientific and other associations, that are so numerous in all American towns, even in those of comparatively small population. The institution of these libraries, both free and otherwise, is an admirable means of diffusing knowledge, and one that cannot be too much admired, especially when conducted on the plan of the Boston Public Library. This institution contains 260,000 volumes and 100,000 pamphlets, which are, together with the use of a good reading-room, free to all ; and residents of the City are allowed to take books home with them.

A visitor to the States must arrive at the conclusion that, as a people, the Americans are more abstemious than ourselves ; for although a great deal of drinking takes place at the numerous bars, yet a drunken man is a rare sight ; and the streets after dark in most of the cities are quiet and orderly, forming in that respect a favourable contrast to those of our own large towns. This is the more remarkable as very little beer is made in the country, except the innocuous lager bier ; and people are therefore forced to drink the wretched Kentucky whiskies. The consumption of wine is very small, but this is no doubt in consequence of its high price ; for strange as it may appear, the cost of the native wines is quite equal to that of the imported; so that what would and ought to be a good and wholesome beverage for the people at large, and a means of keeping them from the pernicious stuff sold under the name of whisky, becomes a luxury available by the rich only.

In glaring contrast to the general cleanliness of the people is the filthy habit of tobacco-chewing and consequent constant expectoration, which is so common, that it may almost be called universal ; although I did not observe that it prevailed to any extent amongst the better class. Hotels and all public places of resort are plentifully supplied with spittoons, but the floor is always covered with discoloured saliva and at first before a stranger becomes inured to the disgusting custom, he will often be seized with a feeling of nausea. It seems strange that the ladies, the greatest sufferers from this filthy habit, do not take some measures to modify it; as at present it prevails to such an extent that no floor, not even that of the Legislative Hall, is free from its traces.

I should say that as a rule English is equally as well spoken in America as in England; for though the nasal twang is very perceptible amongst certain classes in the former, it is not more offensive to the ear, than is the vulgar Cockney pronunciation or the Yorkshire or Somersetshire provincialisms. Educated Americans speak well, with only a slight inflection of the voice at the end of a sentence ; which causes a slight sing-song intonation. Of course many expressions, such as " I guess," are general ; but it is a mistake to suppose, that many of those put in the mouth of the typical stage Yankee, such as " You bet" &c. are so. Even amongst Americans of the lower class, such expressions as " You bet" are looked upon as provincialisms.

I have elsewhere noticed the prevalence of hotel and boarding-house life, and the absence of that "home comfort " so dear to Englishmen ; the want of which acts detrimentally upon the rising generation.

The Americans are a religious people, judging by the number of places of worship in the different cities and the generally strict observance of Sunday, except in a few places where there is a large foreign population. There is no State Church, but the Protestant Episcopal Church corresponding with the Church of England, with its high and low divisions, seems to appeal more to the sympathies of the better class, and now that the pew-rent system which perpetuates class distinctionss in a place of worship, is being gradually done away with, its scope of usefulness will be much enlarged. The Church of Rome as might be expected numbers its adherents amongst the Irish principally. The leading denominations in the States, in point of numbers, are the Methodist and Baptist. It may be mentioned that in the New England States,

Unitarianism has rapidly increased and Evangelical principles have in proportion declined; the former now numbering amongst its votaries many of the most intellectual men in the country. New England too, the land of the Pilgrim Fathers, and where at one time puritanical austerities were enforced by cruel enactments and laws, has given birth to the principal of those religious excrescences of the nineteenth century, that have sprung up with a fungus-like growth until in many cases they count their votaries by hundreds of thousands.

Amongst these out-growths of Puritanism, may be enumerated Mormonism, Universalism, Spiritualism, Materialism, Shakerism, and Free-Loveism; and their rapid growth show how tired the people must have been of religious austerities, when such a re-action could take place as the one that has thrown them by thousands into the arms of these new beliefs.

CHAPTER XX.

POLITICAL CHARACTERISTICS.

POLITICAL Constitution—Effects of Manhood Suffrage—Vote by Ballot—Effects of the Virulence of the Press—Civil Service—Corruption in the Public Service—Venality of Courts of Justice—Reaction—Conflicting Authorities in the State—Difficulty of Central Government—Political Future of the United States.

THE Territory of Colorado having lately been admitted as a State into the Union, the United States now comprise thirty-nine distinct Republics or States, each of which is self-governing under a separate Constitution. It is in fact a league of Sovereign States, banded together for mutual protection and benefit, each of which delegates a portion of its power to a Central Government, legislating on matters affecting the whole.

The division of supremacy between the Union and the States is defined as follows :—"The powers delegated by the Constitution to the Federal Government are few and defined. Those which are to remain in the State Governments are numerous and indefinite. The former will be exercised principally on external objects, as war, peace, negotiation and foreign commerce. The powers reserved to the several States will extend to all the objects which, in the ordinary course of affairs, concern the internal order and prosperity of the State." There are however certain general interests which can only be attended to with advantage, by a general authority, and thus the Central or Federal Government has been further invested with the power of controlling the monetary system, of directing the post-office, of opening the great roads which establish communication between the different parts of the country, of legislating on bankruptcy, of granting patents, and other matters in which its intervention is necessary. Lastly, as it was imperative that the Federal Government should be able to fulfil its engagements, it was endowed with an unlimited power of levying taxes.

The Government of each State is vested in a Governor, elected by the people, and a Legislative Assembly, consisting of a Senate and a House of Representatives. The Senate is generally a legislative body, but it sometimes becomes an executive and judicial one. It assumes executive power in the nomination of public officers, and judicial power, in the trial of certain political offences. The House of Representatives has no share in the administration, and only partakes in the judicial power in so far, that it impeaches public officers before the Senate. The members of the two Houses are elected in the same manner, and by the same electors. Senators, however, retain their seats for a longer time than do the members of the House of Representatives.

The Governor is the chief executive officer in the State, his duties being to lay the wants of the country before the Legislative body ; to point out means to be usefully employed in providing for them ; and to see that the wishes of the people as set forth by their representatives, are carried out. The Governor is head of the Militia and the Commander of the regular forces of the State.

Manhood suffrage prevails in each State, without any qualification, educational or otherwise, except residence in the State.

The constitution of the Central or Federal Government, which sits at Washington differs in little from that of the various State Governments. It consists of the President of the United States, and a Congress, comprising a Senate, and a House of Representatives. The Vice-President is ex-officio President of the Senate, and is elected in the same manner, and at the same time as the President ; that is, by electors appointed for the purpose, by the people of the various States. These electors are in number equal to the members of the two Houses of Legislature of the State they represent, and must not be members of either House. The original intention of the Constitution was, that the candidates for the Presidency should be nominated by these electors, but this is not carried out ; the method of nomination now being as follows :—Conventions are held by the two political parties in the States, called respectively the Republicans and the Democrats, for the purpose of choosing two from amongst the number of candidates for the Presidency and Vice-Presidency; who are put forward as the nominees of the party, and the electors chosen by the people, are pledged to vote for the nominees of that party, by a majority of which he has been elected. The result of this innovation is, that there is not that secrecy, and inviolability, about the election, that was originally aimed at, and for six months before it takes place, there is a political ferment, and excitement, throughout the country, that cannot but be detrimental; as, during that time, legislation may be said to be practically at a standstill.

The President is elected for four years and is eligible for re-election. He can veto any act passed by the two branches of the Legislature, but if passed a second time by a majority of two-thirds, it becomes law.

Senators to the Upper House have the title of Honourable; their term of office is six years, and they are chosen not by the people direct, but by the Legislatures of the various States; two being elected by each State in its sovereign capacity, irrespective of area or population; thus Rhode Island with an area of 1,306 square miles, and a population of under 240,000, returns the same number as New York State, that contains 46,000 square miles, and 4,500,000 inhabitants.

Representatives to the Lower House, who retain their

seats for two years, are elected by the people at large, in the proportion of one, to every 120,000 of the population.

Secretaries of State and other Executive officers are chosen by the President, and have no seat in Congress; so that the opinions of the President and his ministers can only penetrate into Congress indirectly. The President selects persons to fill posts in the civil service, but his nominations have to be approved and ratified by the Senate. The President, in common with all other Executive officers, can be impeached by the House of Representatives before the Senate; and, if convicted, can be debarred from occupying any office in the State.

The entire judicial power of the Federal Government, by which it enforces its laws, is centred in one tribunal, denominated the Supreme Court of the United States. To facilitate the expedition of business, inferior Courts were appended to it, which were empowered to decide causes of small importance without appeal, and with appeal, causes of more magnitude.

The Supreme Court of the United States consists of a Chief Justice and eight Associate Judges, nominated by the President, acting with the consent of the Senate. In order that these judges shall be independent, their office is declared inalienable, and their salary, when once fixed, cannot be altered by the Legislature. The Supreme Court has the power of determining all questions of jurisdiction arising in the States.

Each of the judges of the Supreme Court annually visits a certain portion of the Republic, in order to try the most important causes upon the spot : the Court presided over by this judge is called the Circuit Court. The Union is also divided into districts, to each of which a resident Federal judge is appointed, and the Court which he presides over is termed a District Court.

In addition to these Federal Courts, each State has its own complete judicial system, by means of which obedience is exacted to the State laws, and the life and property of citizens protected.

The theory of the American Constitution is grand in principle, the intention being the government of the people by the people, in such manner, as to attain the greatest happiness of the greatest number. It is sad therefore, that such a large amount of corruption has been allowed to creep into the public service, to render the grand system of Government almost nugatory.

This has been brought about by a variety of causes, foremost amongst which, is Manhood or Universal Suffrage without an educational qualification, the consequences of which are that the votes of that large section of the people corresponding with our middle-class, who have a stake in the country, and a consequent interest in maintaining order, are swamped by those of the large uneducated Irish-immigrant class and of the Negroes. The better class, consisting for the most part of born

Americans of good education and means, finding themselves thus outvoted, withdraw entirely from political affairs.   As is usual too, when the franchise is extended to an uneducated class, the people composing it, are led by demagogues and stump orators; and the use of their voting-power is not regarded by them in the light of a duty devolving upon every citizen of a State, to be conscientiously performed, but is looked upon solely as possessing a marketable value, and is sold accordingly, in the same manner as any other of their possessions. Thus election frauds are of such common occurrence in America, and the system of vote-by-ballot, if anything, tends to increase them.

The system of vote-by-ballot was instituted to ensure secrecy of election, but the mode in which it is carried out causes it to fail in attaining this object.   Two differently-coloured voting-papers are used, one for the Republican candidates and the other for the Democratic, denominated the party " ticket;" hence the expression " voting for a particular ticket."   The voter strikes out the names of the candidates he does not wish to vote for; gives in his name, which is checked by the Returning Officer, from a list of the registered voters; and deposits his "ticket" in the ballot-box, open, to prevent the possibility of two being folded together.   It will be seen, that this voting system is anything but secret, and consequently no protection against bribery.

It is now nineteen years since a committee reported to the Assembly of New York State that the ballot "still fails to be a true reflection of the will of the people," and since then, things would seem to have remained in *statu quo*, in spite of the registry laws enacted by the Legislature of that State in 1865; for the following were enumerated, by a committee appointed to enquire into the conduct of the elections of 1868, as being the most prominent frauds in connection with it perpetrated in the City and State of New York :—

" 1. Many thousands of aliens fraudulently procured, or were furnished with certificates of naturalization illegally or fraudulently issued, by means of which they were enabled to register as voters and voted in violation of law."

" 2. Many hundreds of certificates of naturalization were granted in the names of fictitious persons, to be used by native-born and naturalized citizens and aliens in falsely registering as voters, and to enable them to vote many times at the election."

" 3. Many hundreds of persons voted in New York City from two to forty times or more, each under assumed or fictitious names fraudulently registered for the purpose."

" 4. Extensive frauds were committed in canvassing tickets, and names of voters were registered on the poll-lists, and democratic tickets counted as if voters representing them voted, when no such persons voted at all."

" 5. To accomplish these frauds, gross neglect of duty and disregard of law so great as to evince a criminal purpose prevailed in some of the courts, while officers and democratic partizans of almost every grade, either by official influence or otherwise, aided, sanctioned, or knew of and failed to prevent them. The same influences shielded the perpetrators in nearly all cases from detection or arrest, and when arrested they have, through the agency of judicial officers and others charged with the duty of prosecution, escaped all punishment."

" 6. Through these agencies the democratic electors of President and Vice-President and the democratic candidate for governor of the State of New York were fraudulently elected."

" 7. And the investigations of the committee show that existing State laws and the mode of enforcing them are wholly inadequate to prevent these frauds, but that Congress has the power to enact laws which, if faithfully executed, will, to some extent furnish remedies hereafter."

" There is no law of Congress professing to prevent or punish frauds in voting or conducting elections ; and the penalties relating to certificates of naturalization are by no means adequate."

The above excerpt from the State paper refers to frauds in connection with the State of New York, but the report states that Maryland, Louisiana, and other States have presented phases of the same evil. It must also be borne in mind, that although the above report refers to Democratic frauds, that the Republicans are charged by their opponents with like mal-practices, and it is generally conceded that they are tarred with the same brush.

The press acts very prejudicially against the public service of the country, and the effects of its virulence in respect to public men, are seen in the fact, that the business of politics has sunk in public estimation, and men of the better class keep aloof from it. All offices in the State consequently fall into the hands of second-class men, to whom the salary and perquisites of office are of vital importance.

The Civil Service of the country is not a fixed one, and all office holders in it retire at every new Presidential election ; the vacant posts being filled by the supporters and friends of the new President. It must be apparent, that this cannot be conducive to the proper conduct of the various State Departments ; for as a new Presidential Election takes place every four years, a person appointed to an office under government has just time to learn the routine of his work, when he has to retire, and make room for a successor, who has no knowledge of the duties he is about to undertake. To this may be attributed in a great measure the frightful extravagance that pervades every State Department and as far as my own experience of one of them—the Post Office—goes, I can only say if the others are not better managed, then they are a disgrace to such a country as the United States ; for I lost more letters during my short stay in America, than I have ever done before, and the difficulty I experienced in obtaining foreign letters addressed to me at the Post Office, was simply atrocious. On one occasion the mail had arrived from Australia, and expecting letters, I went to the Post Office after the mail had been delivered, and friends of mine were already in possession of their communications. After answering the enquiry whence I expected my letters, I was told there were none for me, but not satisfied with this assurance I again went in the afternoon and received the same reply. Insisting however that there must be one or more for me I got the clerk again to look, and sure enough this time he managed to find three for me, two that had arrived by the Australian mail, and one by the English. At that very time there must have been two more letters of mine lying in the office, but they never came to hand. Certainly I would not like to say without more data, that the other Departments are managed in the same manner, but in any case it cannot be desirable, that the President should have such an undue amount of patronage in his gift as the nomination to forty thousand offices in the State, as is the case at the present time.

I have attempted to point out the various causes that have served to bring about the present corrupt state of affairs in the United States, and I will now mention what, from observation I found that state of affairs to be.

From one end of the country to the other complaints are general of the corruption that pervades every Government Department, and everybody with whom I came in contact seemed persuaded that votes are bought

and sold; that gross bribery and other mal-practices
prevail at all elections; that the very fountains of justice
are polluted and judgments sold to the highest bidder.
Now these complaints are not confined to a class or
section of the people, but are universal, being attributed
by the Republicans to the Democrats, and *vice-versâ*.
Certain it is that the maxim prevails "to the victors
belong the spoils," and if say, a Republican President be
elected, then all the offices in the State are filled by his
own supporters, who, knowing that they are only in power
for four years, are consequently less able to resist the temp-
tation to "feather their nests" in that time.   More especially so, since it seems to be considered part of the existing
political creed, that as office holders are appointed by
the President the representative of the majority, from the
majority, they are justified in enriching themselves at the
expense of the minority.   Thus it is that Americans
distrust their public men, and impute to them dishonesty,
as a matter of course, without pausing to inquire whether
there be any grounds for such a charge.

The same accusations of venality extend to those
Courts of Justice the Judges of which are appointed by
the State Legislature, and it is universally believed by
Americans that, if a criminal only be a wealthy man, it is
almost a matter of impossibility to obtain a conviction
against him.   It is not the Judges alone who are accused
of being venal, but Juries in the States are supposed to

be particularly open to convincing proofs of innocence, when conveyed in that peculiar chinking metallic sound that carries with it such weight and conviction. Persons have been pointed out to me, in various cities in the States, walking about at large (in consequence of their having notoriously bribed the persons appointed to try them), who, at some time or other, had been on their trial for offences committed, concerning whose criminality no moral doubt could exist, and who, in any other civilized country in the world, would have been working out a sentence to a long term of imprisonment.

It seemed to me, however, that a great re-action was taking place in public opinion, for there seemed to be a general desire for reform, and the Conventions recently held at Cincinnati and St. Louis for the purposes of choosing the Republican and Democratic candidates for the Presidency and Vice-Presidency, have nominated good men in both cases. It is generally hoped whichever party succeeds in getting its candidate elected, that the new President will have the moral courage to sever himself from the traditions of his party, and to devote himself to the bringing about of the much desired reforms in the public service. He will have an opportunity of gaining the goodwill of his fellow-citizens ; of making for himself a niche in history, and of having his name handed down to posterity with that of Lincoln as a benefactor of his country. Sincere well-wishers of the great Republic, and those who believe in its great destinies, must hope that this reform movement is not a temporary excitement incidental to a Presidential election, that may wear away and allow affairs to remain in *statu quo*.

The weakness of the constitution of the United States seems to be in the existence of two conflicting authorities in the government, viz., the Central or Federal, and the various State Legislatures ; for, as each State in the Union is self-governing, and the powers of the central authority are prescribed within somewhat narrow limits, the latter cannot be strong, nor have the same power as other national governments, that do not possess a divided authority in the State. Nor is it advisable for the maintenance of the Union, that such should be the case, except for defence against foreign aggression ; for the country has increased so rapidly that it now covers an area of more than three and a half millions of square miles, that is, a larger area than the whole of Europe.

Such an immense tract of country, with its diversity of climate, naturally produces conflicting interests : thus it is easy to see that the interests of the North, the great manufacturing centre, would be different to those of the South, the resources of which consist in its cotton, tobacco, and other products of a tropical or semi-tropical zone ; that these again would be opposed to the interests of the great West, the grain-producing States, and still more widely diverge from those of the Pacific States, whose principal resources lie in their great mineral wealth. Now the difficulty experienced by a Central Government in legislating for such antagonistic interests is abundantly shown in the working of the protective tariff imposed by the Federal Government for revenue purposes, and also as a means of paying off the National Debt contracted during the late civil war. The effects of this measure, are to increase the cost of all manufactured articles throughout the country, and more especially to benefit the North, at the expense of the West and South. It must be plain therefore that were the powers of the Federal Government enlarged and it had to enact laws, say for the sale of the land, throughout the country, that similar results would be produced, as has been the case with the tariff; that is, some of the States would be benefited at the expense of the others. It would seem therefore, that the continuance of the Union depends upon, and its interests are best served by, maintaining the Sovereignty of the various States.

The Republican or as it might be called the Radical party which comprises the people in the New England and Middle States, and a large proportion of those in the Western appears however, to aim at centralization; that is, the formation of an undivided Republic governed from Washington, and with this end in view continually overrides the Constitution in its attempts to subvert State

Independence. This was the cause of the late civil war ; for the Southerners fought for State Rights, including the power to secede from the Union; which the Northerners wished to maintain in its integrity. It does not require a prophet to foretell that if the system of centralization be extended, as it undoubtedly will be if the Republican party remain in power, that the civil war will be repeated, and will end more disastrously for the Union than the last ; for as it required the combined power of the Northern and Western States, to put down the rising in the South, it may easily be conceived that it would be quite impossible for the Northern States alone successfully to combat the Western, with the moral support they would have from the Southern and Pacific States.

At present such an eventuality is not thought of, but I have heard many enlightened Americans express their abhorrence of the existing state of affairs, which does not properly protect life and property; which cannot prevent justice from being bought and sold ; and which allows corruption to creep like a cancer into the public service. I have heard too, many declare their preference for the tyranny of one, rather than of a number, and express their opinion, that an absolute Imperialism would even be preferable to the evils of party domination.

Centralization would I believe be the death-knell of the Union, and taking into consideration the vast size of the country and what the population at its present rate of increase will amount to, say twenty years hence, it is a moot point whether the interests of the people would not be better served, and civilization generally be more benefited, by the formation of three distinct Governments —say a Northern, a Southern, and a Western.

There seems however to be no reason why the Union should not be maintained, if State Sovereignty be properly recognized; especially if the better class of the people would take a more prominent part in the government, and when the evils of Universal Suffrage are modified by the effects of the diffusion of education.

THE END.

Walker, May, and Co., Printers, 9 *Mackillop-street, Melbourne.*

22